North Carolina Weather & Climate

Published
in association
with the State
Climate Office of
North Carolina by
The University of
North Carolina Press
CHAPEL HILL

Peter J. Robinson

North Carolina Weather & Climate

Designed by Heidi Perov
Set in ITC Charter and The Sans
Manufactured in Korea

The paper in this book meets the guidelines for permanence and durability of the Committee on Production Guidelines for Book Longevity of the Council on Library Resources.

Library of Congress Cataloging-in-Publication Data

Robinson, Peter J.
North Carolina weather and climate / by Peter J. Robinson.
p. cm.
Includes bibliographical references and index.
ISBN 0-8078-2961-7 (cloth: alk. paper)
ISBN 0-8078-5625-8 (pbk.: alk. paper)
1. North Carolina—Climate—Popular works.
2. Meteorology—North Carolina—Popular works. I. Title.
QC984.N8R63 2005
551.69756—dc22 2005008003

Publication of this work was aided by a generous grant from Capitol Broadcasting Company, Inc.

cloth 09 08 07 06 05 5 4 3 2 1
paper 09 08 07 06 05 5 4 3 2 1

for Shirley,

who has experienced all types of weather

with me and asked all the hard questions

contents

foreword

When I first came to the South in 1981, two things troubled me. First, I wondered if I'd be here for longer than a year (I was on my third job in two years after graduating in meteorology from Penn State). Second, I wondered if I'd ever see snow again (the South to me meant palm trees and temperatures that seldom got below 50°F).

I could hardly have been further off, on both counts. I've found a home in North Carolina and with WRAL-TV. By the time this book comes out, I'll be approaching my twenty-fifth anniversary with the station. And I had come to a state that not only received several snowstorms in a typical winter but that experienced all manner of storms and extremes of weather—more than enough to keep this self-professed "weather junkie" interested and occupied.

For example, the mercury plunged to 4°F on January 11, 1982 (the winter after I arrived), setting the stage for a two-day onslaught of ice and snow. The culmination of this winter weather event was five inches of snow in as many hours! But that five-inch snow was nothing compared with January 2000. What looked to be a nuisance snow event the day before turned out to be the single heaviest snowfall in the history of Raleigh. A tad over twenty inches fell in less than twenty-four hours, leaving many to ask, "How did Greg miss that one?" (Ahem.)

An ice storm, which seems to be a North Carolina specialty, took our station off the air for a time. In December 1989, severe icing occurred on our 2,000-foot transmission tower. The next day, when the sun rose, ice began to melt at a disproportionate rate on one side of the tower, creating an imbalance that led to the complete collapse of our tower. A couple of weeks later, up to twenty inches of snow fell along the southern coast of North Carolina—the southern coast! One more snow note: in May 1992, nearly sixty inches of heavy, wet snow fell in our mountains at Mt. Pisgah.

Of course, having the tallest mountains east of the Mississippi River brings its own extremes. Does it ever get really cold here? How does -34°F hit you? That was the temperature at the top of Mt. Mitchell in January 1985. Admittedly, we're talking about an elevation of more than 6,600 feet above sea level. Well, how does -9°F in Raleigh sound? It really happened, on the very same morning, crushing the all-time low temperature record set in February 1899 by seven degrees!

But before I give you the wrong impression, North Carolina is not the Montreal of the South. Yes, it gets hot here—man, does it get hot! Fayetteville set the state record for heat with a sizzling 110°F in August 1983. Extended periods of heat and drought are not uncommon in North Carolina. Relief from those droughts often arrives in the form of tropical systems during the late summer and early fall. North Carolina has a storied history when it comes to hurricanes, with hurricane Hazel in 1954 being labeled the benchmark for many years. Our state enjoyed a relative vacation from serious hurricanes during the 1970s, 1980s, and the first half of the 1990s. But boy did things change in 1996. Hurricane Bertha roared ashore in July as a strong category 2 storm just twelve hours after it appeared to be on its last leg. Then in September, hurricane Fran arrived as a category 3. This tropical troublemaker maintained much of its strength as it moved inland toward Raleigh, leaving what appeared to be a war zone in its wake. In September 1999, hurricane Dennis passed just offshore, stalled, and eventually made landfall as a strong tropical storm five days later. Dennis dumped five to ten inches of rain across the eastern half of North Carolina. Less than two weeks later, a supersized hurricane Floyd dumped up to twenty inches of rain on the same areas soaked by Dennis, and the result was the worst flooding in recorded history. And we can't forget the most violent storm on the face of the earth—the tornado. Yes, North Carolina experiences its fair share of these powerhouses as well. Two of the more notable events occurred in March 1984 and November 1988. The former killed 57 people and injured more than 1,200 in both North and South Carolina. Then in 1988, as the Thanksgiving weekend was winding down, an intense tornado developed at 1:00 A.M. near the Raleigh-Durham Airport. This monster traveled on the ground more than eighty miles during the next two hours. Miraculously, only four people lost their lives, but the damage was severe and life changing.

So North Carolina has plenty of weather, and it comes at us from all directions of the compass. We have easy access to the cold of Canada, the warmth of the tropics, and an abundance of moisture from both the Gulf of Mexico and the Atlantic Ocean. The proximity of the Appalachian Mountains to the warm Gulf Stream waters can produce an incredible variety of both temperature and precipitation types across our state, sometimes over just a few miles! As you can tell, I remain excited about North Carolina's climate, but the words you have just read only scratch the surface of our exhilarating and sometimes extreme climate. So, do you want the rest of the story? If you do, there is perhaps no one better suited to fill in the gaps than Peter Robinson. Peter served North Carolina as its state climatologist in the 1970s and has been a professor of geography at the University of North Carolina at Chapel Hill for more than thirty years. He has accumulated a vast amount of knowledge about our climate and recognized that no one had ever published a comprehensive summary of North Carolina's extremely dynamic climate. Because of his interest in global climate change, Peter is uniquely qualified to perform an objective evaluation as to how North Carolina's climate changes fit into the global picture. If you are a newcomer to North Carolina, this book will help you to see immediately just how diverse our climate is. And even if you're a native of our state, you may not realize how much meteorological madness you can find here! In his own unique way, Peter explains for you all the variables that go into making our weather among the most exciting in the world. While Peter is an academician, he knows how to write in such a way that anyone with even the remotest interest in weather and climate will find it next to impossible to put this book down. You might even learn some basic meteorology! So, sit back and enjoy the ride. Peter Robinson is about to take you into the exciting world of North Carolina's climate. Yes, this is a climate that I once feared would be mundane. The last quarter-century has shown me otherwise.

—Greg Fishel

preface

For more than thirty years I have been involved with North Carolina's climate—professionally as teacher and researcher but more personally as one who has lived in it. Coming from Britain, I found the range, variety, and sometimes violence of North Carolina's weather intriguing, challenging, and stimulating. Camping in all kinds of weather, including abandoning a campsite at Cape Hatteras as an offshore hurricane got closer, throwing away a sodden tent after a week of rain in the Smokies, and watching glorious sunrises or sunsets from the mountains to the sea, always raised questions about why our weather behaves the way it does.

This interest took a practical form during my stint as the state climatologist of North Carolina. Inquiries by citizens of the state, whether they were individuals planning gardens or simply wanting to know something, or large corporations needing information to help in their operations, provided a never-ending series of questions. Some were easy to answer, some took much thought and effort, and some I could never answer. Indeed, these questions and questioners are the real source of this book, and I am grateful for them all for providing a wonderful opportunity to learn more, and enjoy more, about our weather over the years. Providing answers—or trying to provide answers—has been tremendous fun.

The State Climate Office of North Carolina, now located at North Carolina State University, has been through the years a constant source of help, inspiration, and information. As a partner in this book, the office has provided assistance beyond measure. Without the support of State Climatologist Sethu Raman and the immense effort of Ryan Boyles, the Associate State Climatologist, this book would never have appeared. My grateful thanks to them.

Many other people have, consciously or not, contributed to this book. At the front line are my students, whose looks of incomprehension led—eventually—to many of the explanations given here. Colleagues in the geography department, especially Rich Kopec, Art Dodd, and Chip Konrad, have always been ready and willing to discuss any aspect of the weather. The same can be said for colleagues in other departments at the University of North Carolina at Chapel Hill and North Carolina State University, at the Environmental Protection Agency in the Research Triangle, and in the local National Weather Service Offices in the state. Friends on other UNC campuses, mainly in geography departments, have vastly increased my understanding of local conditions from the coast to the mountains. Further, we are fortunate that the National Climatic Data Center is in North Carolina. Especially in the days before instant Internet access, a visit to them was not only an excuse to visit Asheville (especially in the summer) but a chance to talk to fellow climatologists, and even to collect some data. Finally, one group especially, the Central North Carolina Chapter of the American Meteorological Society, has over the years fostered a relationship among researchers, teachers, operational meteorologists, media people, and consultants that has created a knowledgeable and enthusiastic association that has not only benefited me personally but been invaluable to meteorology and climatology in this state. With all these people, too numerous to name individually, giving advice and information over the years, there should be no errors in this book. But I am sure that there are. They are my fault.

North Carolina Weather & Climate

chapter 1

Weather in Our Lives

North Carolina weather is wonderful in its variety, from the quiet of a cloudless mountain morning after a winter snowstorm to the splendor of dogwoods blooming on a bright spring day in the Piedmont, and even to the tempest of a hurricane approaching the coast on an autumn night. Knowing more about the weather helps me appreciate it more and live with it more wisely. I hope this book helps you do the same.

The weather of North Carolina is ever changing. Most of the time, fortunately for us, it is favorable and allows our crops to grow prolifically, provides the water resources we need, and creates the buds and flowers of spring, the vivid colors of autumn, and the recreational opportunities of summer and winter. Sometimes, of course, as a hurricane bears down, a tornado roars through, or floods loom upstream, we are reminded of the less friendly face of nature. The constantly changing nature of our atmosphere makes forecasting the weather a challenge. For a few days the weather may do exactly what we expect of it. In April we may predict, and get, a week with balmy spring days. But then suddenly winter returns. Sometimes we forecast the change; sometimes it seems to arrive out of nowhere.

In this book we will look at what our North Carolina weather is and why it is that way. We will mix specific information about our weather and climate (using pictures, tables, and graphs) with explanations about its causes (using words and diagrams). In each chapter some particular, usually extreme, weather events are highlighted. The more we understand how the weather works, the better we can forecast it. So we will use our understanding to discuss forecasts, not only those made by professionals but also those we can make ourselves. We will look at how weather affects us in our daily lives and how we can use our knowledge of its ways for our own benefit—whether at work or at play. We will consider some of the issues facing us in our state as the weather varies from day to day and from year to year, and we will examine the problems or threats posed by air pollution and global climate change.

Some North Carolina Weather Basics

Most of the time our weather comes from a generally westerly direction, so if we want to give our own short-term weather forecast, we can usually look to the west to see what is being "blown in" for us. This common wind direction is not just happenstance but is dictated by North Carolina's location on the globe. Our state, positioned between 34°N and 36.5°N latitude, is near the southern edge of a great river of westerly flowing air between about 30°N and 60°N. This is the region of the *westerlies* (see Fig. 1.1). Like a river, this air does not flow smoothly, and the main current meanders about a lot, giving us the daily weather sequences. Slower, bigger changes largely create our seasons. In this section we set the stage for considering all of these facets of our weather.

WEATHER THROUGH THE YEAR

In winter the main current of air is often nearly overhead. The result is a series of storms that come from the west, with plenty of rain or snow. The airflow is also cold, particularly if it has come from the northwest rather than directly from the west. Indeed, the source of this frigid air is often the Canadian Arctic. This northwesterly airflow is cold for two reasons. The first is the geometrical relationship between the sun, which provides the energy to warm our planet, and the earth itself. The closer to the pole you get, the less energy you receive from the sun, and the colder is the air. So the air coming to us from the northwest is cold. The second factor involves complex links between the sun's energy, the underlying ground, and the overlying air. In winter, land surfaces tend to cool more, and they reach a lower temperature than do water surfaces at the same latitude. The opposite happens in summer. So in winter the air coming from the northwest is especially cold because it has been blowing for a long distance over a cold continent. If that surface is also snow-covered, we are likely to get even lower temperatures. So this airflow from the west or northwest gives us cold, clear, and dry weather. Some small clouds may occur, especially when air has to rise to get over our western mountains (see Fig. 1.2). At other times in winter we are dominated by air blowing in from the south Atlantic Ocean or the Gulf of Mexico, which brings a very different type of weather: warmer, more moist, and more cloudy. Day-to-day weather

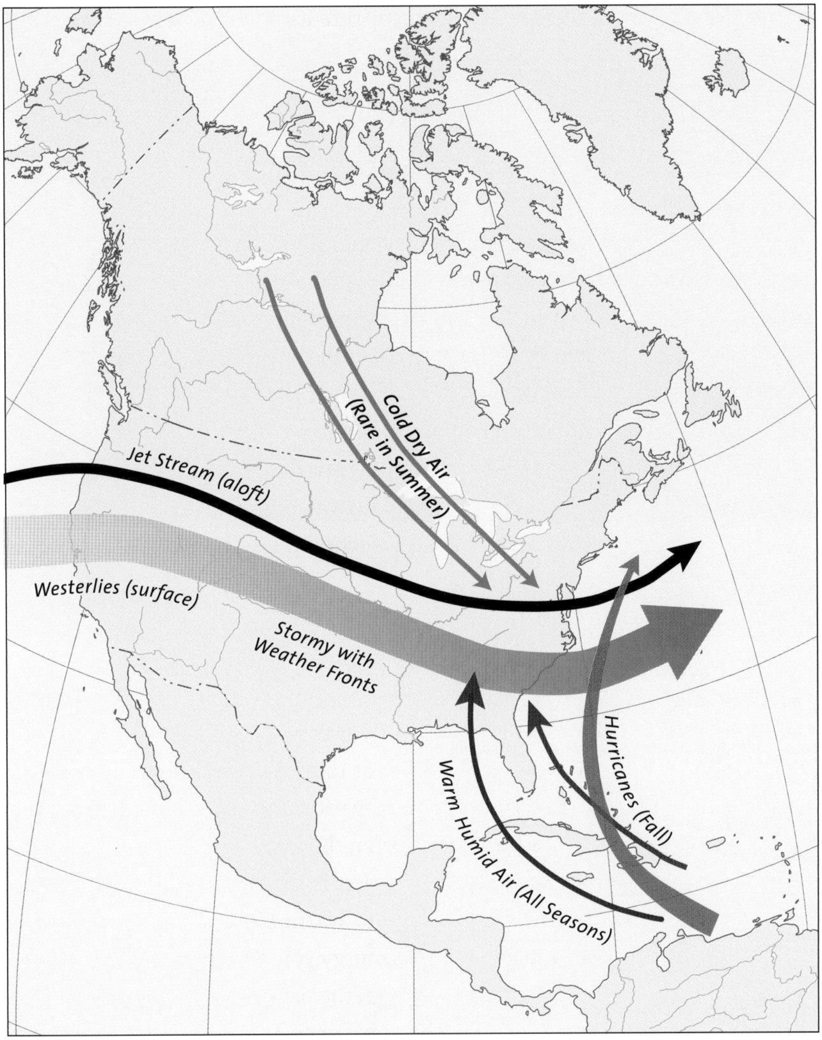

FIGURE 1.1. *Schematic view of the major features controlling the weather in North Carolina*

changes are a marked feature of our winter weather and climate. Throughout it all, however, we expect the mountains to be colder than the eastern part of the state, simply because temperatures usually decrease with height. Also, following from this, we usually get more snow in the mountains. But this is not always the case (see Box 1.1).

In summer the main current of the westerly airstream is usually some distance to the north of our state. We are close to the edge of the flow, and conditions are often nearly calm. Indeed, wind speeds are lower in summer than in winter. For much of the summer, air drifts into the state from a southerly direction. This

FIGURE 1.2. *Air flowing from the west over the mountains into our state is forced to rise. In this case the air has sufficient moisture and uplift to form a thin layer of clouds. At other times deep clouds may be created.*

air has been over the warm waters of the Atlantic Ocean or the Gulf of Mexico, and lots of that water has evaporated into it; so it is no surprise that humid air dominates. Thunderstorms can generate readily in this rather calm, moist area, and they are a major source of rain in summer. They are often responsible for the intense, short-lived, and local showers of a summer afternoon. At other times in the warm season, the main current of the westerly flow may swing south, pick up moist air from the Gulf of Mexico, and sweep across our landscape as a series of weather fronts and storms that bring periods of more widespread and more gentle rain. Although the type and intensity of the rainfall may vary depending on the direction of the airflow, all of these airstreams give a similar type of weather: hot, humid, and frequently cloudy. Temperatures remain high throughout the season, and there is much less variability in the day-to-day weather in summer than there is in winter.

This description refers to our weather as it commonly occurs. Many things may disrupt this general pattern. Spring and fall are transition seasons and often mix the characteristics of summer and winter to give their distinct weather. As summer turns to fall, the ocean gets hotter, and evaporation off the tropical Atlantic Ocean gets more vigorous. The conditions that produce summer thunderstorms become ripe for hurricane development (see Fig. 1.3). Some years can pass without significant hurricane activity; then conditions change slightly, and a storm comes along, bringing high winds, dumping a tremendous amount of rain, and causing major disruption and damage. The weather history of the state suggests that most

BOX 1.1. THE CHANCE OF A WHITE CHRISTMAS AND THE STORM OF 1989

Christmas comes rather early in the winter, and so the chances of snow on that day are low everywhere in North Carolina. For most of the state the odds are less than one in twenty. In our mountain towns it rises to about one in ten. It is more frequent on the mountaintops, but we have few recorded observations at such places.

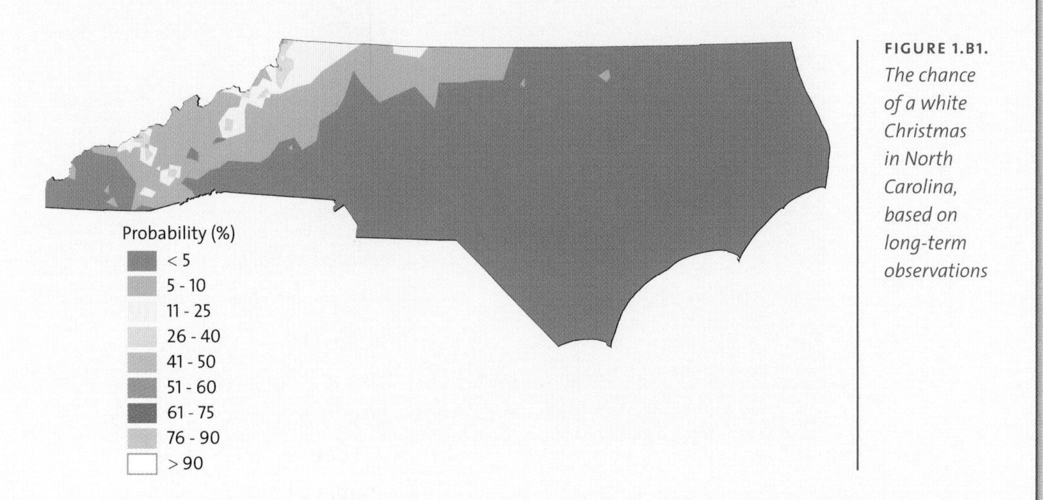

FIGURE 1.B1. *The chance of a white Christmas in North Carolina, based on long-term observations*

The odds may favor snow in the mountains at Christmas, but individual years vary greatly. In 1989 our northern mountains were snow-free until they had a light dusting late on Christmas Day. On the coast the weather was very different. There a storm system moving northward brought snow that started in the evening of December 22 and did not stop until after midday on December 24. Christmas morning saw snowdrifts up to eight feet deep along much of the southern coast! This was the first white Christmas there since at least 1890, when records were first kept in the area.

hurricanes affect the coast but that no place is immune. It also suggests that hurricanes often come in pairs and are often associated with major flooding. Dennis and Floyd in 1999 were clear, but not unique, examples (Box 1.2). The opposite condition, drought, is also a likely visitor for us. Less spectacular, but equally problematic, it may silently affect the whole state for months or even years. Floods and droughts are different expressions of the global pattern of the atmosphere. In addition, all places in our state generate their own local weather and climate. Examples range from the coast creating the sea breezes of a summer afternoon to a mountain valley producing a pocket of frost on a winter morning.

FIGURE 1.3. *Hurricane Fran approaches the North Carolina coast, September 1996 (image courtesy of NOAA/National Climatic Data Center)*

WEATHER AND CLIMATE: IS THERE A DIFFERENCE?

The weather changes from day to day, or even from minute to minute. It also changes with the seasons. We have all observed these changes and readily recognize, for example, that the temperature variations within a typical day in summer are similar to, but not the same as, those of winter. We can think of the hourly changes as being superimposed on the daily ones, which are in turn superimposed on the seasonal ones. Our records also indicate that such superimpositions extend through the decades and seem to work on a millennial scale, or even longer. So fluctuations are piled upon fluctuations, and we can say that the atmosphere operates on a variety of timescales.

The same kind of superimposition also holds when we consider weather over an area. Summer throughout our state is likely to be hot and humid, but superimposed on this general picture are regional variations. We expect the mountains to be cooler than the Piedmont, which in turn is cooler than the Coastal Plain. Even within these regional differences are smaller variations covering smaller areas. Differences in the type of land surface, for example, cause weather differences. So cities are usually warmer than rural areas, swampy regions are cooler than dry agricultural fields, and these differences hold whatever the season. The weather can vary even from one side of a house to the other.

When we start to describe, understand, and forecast weather, we find that there are combinations of the various scales of time and area that are especially useful. The most common combination, and usually the most useful on a day-to-day basis, is the one that gives the *synoptic* forecast. This is the one seen on the daily TV weather forecast (see Fig. 1.4); it offers a prediction for a day or a few days in advance for a major region of the state. The word "synoptic" has the same meaning here as in the "synoptic gospels" of the biblical New Testament: a continuous, and hopefully complete, sequence of events. The forecaster has to tell the day-to-day "weather story" in an unbroken sequence looking at the events of the recent past, the current weather, and the forecast for the days ahead.

The synoptic forecast deals not only with the general conditions a few days ahead but also with short-lived and local features, whether tornadoes or ice storms. It is concerned with the weather. The slower changes, lasting for months or longer and covering a broad area, have traditionally been associated with climate. Historically two very different sets of analysis techniques were needed for them, and two distinct sciences, meteorology and climatology, developed.

In the past, climate was seen as an aspect of the atmosphere that changed very slowly, and it was assumed that if we averaged observations over a long period, we would identify the "normal" climate. By international agreement, thirty-year averages were used to create the *climatological normal*. To determine this value, the last three complete decades of record are used. At present this is the 1971–2000 period. The concept of a long-term average has many practical uses, but the term "normal" is perhaps unfortunate, since there is nothing particularly special or unique about the average over the last thirty years. It is probably best to think of the "climatological normal" as a useful technical term with a special meaning and ignore the fact that the word "normal" has a more common usage.

Our increased understanding of climate has shown that climate changes in much the same way as the daily weather. The main difference between weather and climate lies in the timescale and areas involved. So the techniques used to understand weather and climate separately have been getting more similar, and the methods used to forecast them now overlap so that it is difficult to define the boundary between them. Further, the names of the two sciences themselves these days are often lumped together as "atmospheric science." At least this gets away from any suggestion that meteorologists study meteors—which is a historical holdover from the days when we believed that meteors essentially controlled the weather.

BOX 1.2. DENNIS AND FLOYD, 1999: TRACKS AND RAINFALL

Hurricanes often arrive over North Carolina in pairs, commonly bringing not only wind damage but also tremendous amounts of rain that lead to major floods. Dennis and Floyd in September 1999 were one such pair.

Dennis first came close to the North Carolina coast on August 30 (fig. 1.B2a), with wind gusts of hurricane force and storm surges causing tides from eight to ten feet above normal. The storm then spent four days meandering offshore in a seemingly random way. All forecasters, amateur or professional, were, to say the least, confused. Finally on September 4 Dennis moved inland, more or less on a straight track over the Morehead City region. By then it had weakened considerably, and winds were somewhat below hurricane force. It continued to lose intensity and eventually crossed into Virginia as a weak tropical depression. Total rainfall was not excessive for an active hurricane, with a maximum near 10" on the Outer Banks near Ocracoke, decreasing to about 5" on the eastern edge of the Piedmont (fig. 1.B2b). The rains saturated the soils of the Coastal Plain, and most of the rivers were bank-full. Some places had minor flooding, but there were no major floods after Dennis.

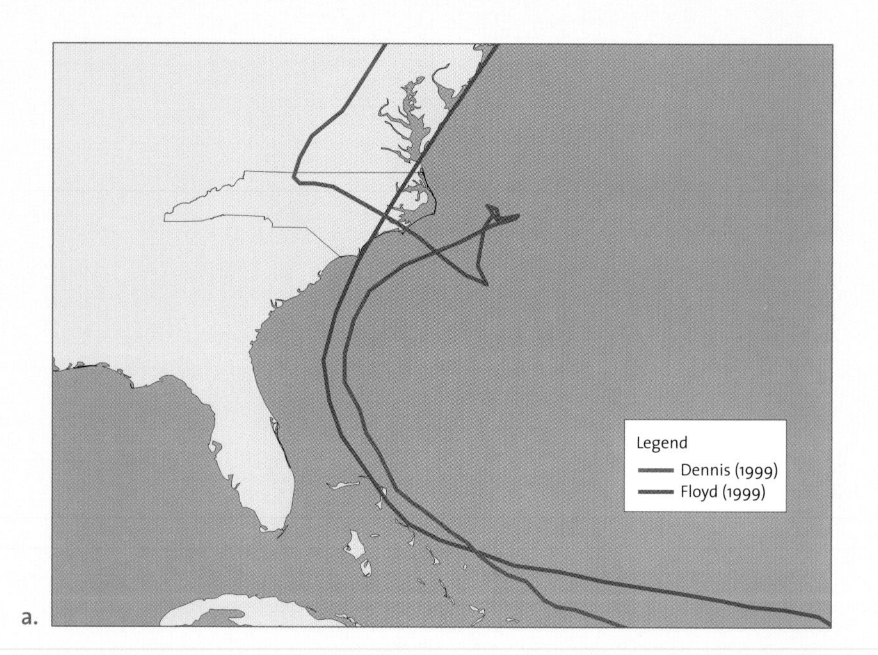

a.

Floyd, the second storm, swept in from the south just two weeks later; its effects were first felt on our southeastern coast on September 14. At that time it was an intense category 4 storm (on the hurricane scale from 1 to 5) off the Georgia coast. It approached us on a track much simpler, and more predictable, than that of Dennis and made landfall near Cape Fear early on September 16. At that time it was a category 2 storm with winds just over 100 miles per hour. It was also fast-moving, and it rapidly passed over the eastern Coastal Plain and into Virginia (fig. 1.B2a). By the evening of September 16 it was over New Jersey. During this short time, however, there was a great amount of rain, approaching twenty inches in places (fig. 1.B2c). This amount of water falling over a broad area onto a surface that was already soaked led to severe flooding in almost all the rivers of the Coastal Plain.

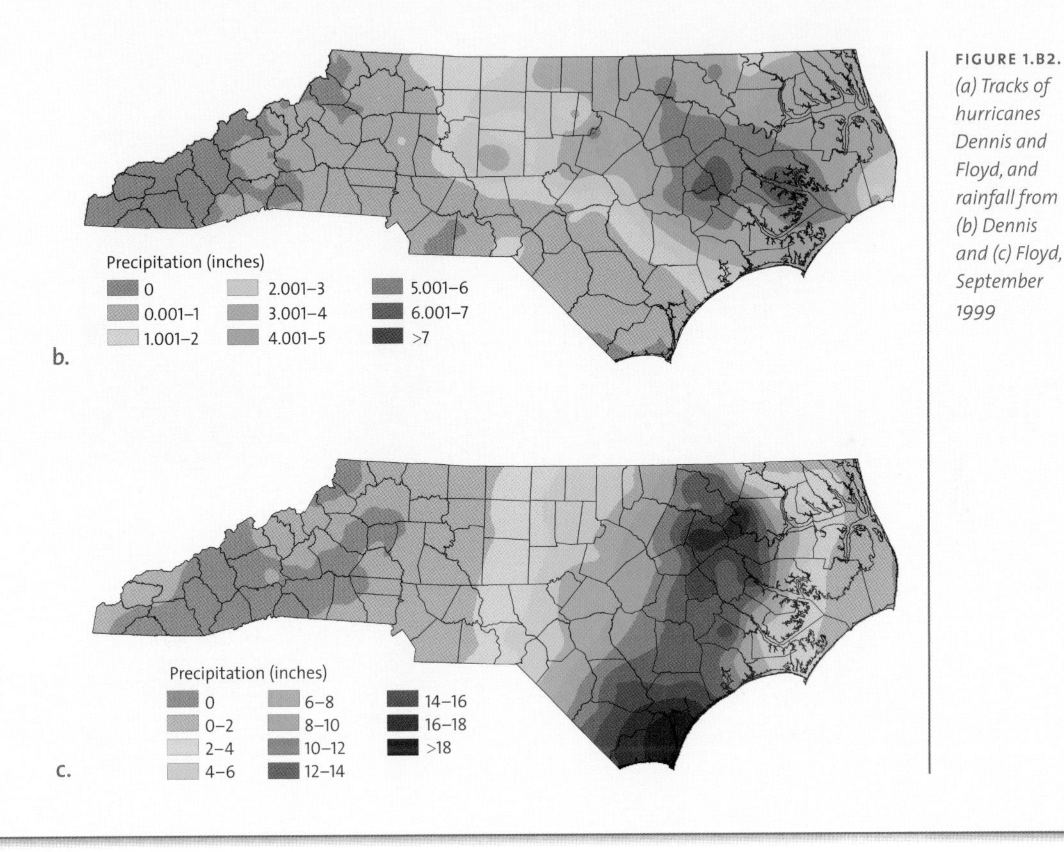

FIGURE 1.B2. *(a) Tracks of hurricanes Dennis and Floyd, and rainfall from (b) Dennis and (c) Floyd, September 1999*

FIGURE 1.4. *Typical daily weather map in a format similar to that seen on many TV shows and websites. This one shows a warm airstream sweeping around a high pressure area near the New Jersey/New England coast and crossing North Carolina from the southwest. A cold front is approaching from the west but is still far distant and unlikely to affect our weather in the next couple of days. (image courtesy of WRAL-TV)*

Even with these recent advances, some differences are retained. We still talk about the temperatures on a particular day being "five degrees above normal," where the normal is the thirty-year average—the climatological normal—for that day. This is, in fact, very useful, since we humans, physiologically and socially, tend to be adapted to conditions that are not too far from normal for a particular region and season. So it is still useful to draw a somewhat vague distinction, with weather as short term and climate as long term. We shall use this division here.

WHAT IS THE WEATHER?

Both weather and climate involve a suite of elements: temperature, sunshine, humidity, precipitation type and amount, wind speed and direction, and cloud type and amount. Sometimes we are concerned with only one element; we often ask, "Will it rain today?" We may even be concerned with only one aspect of an element, such as days with heavy rain (see Fig. 1.5). At other times we may be interested in combinations of elements. Discomfort or even serious health problems can arise from the combination of high temperatures, high humidity, and abundant sunshine in the summer, or the combination of low temperatures and high winds in winter. Many weather events, such as a hurricane, involve virtually all of the elements.

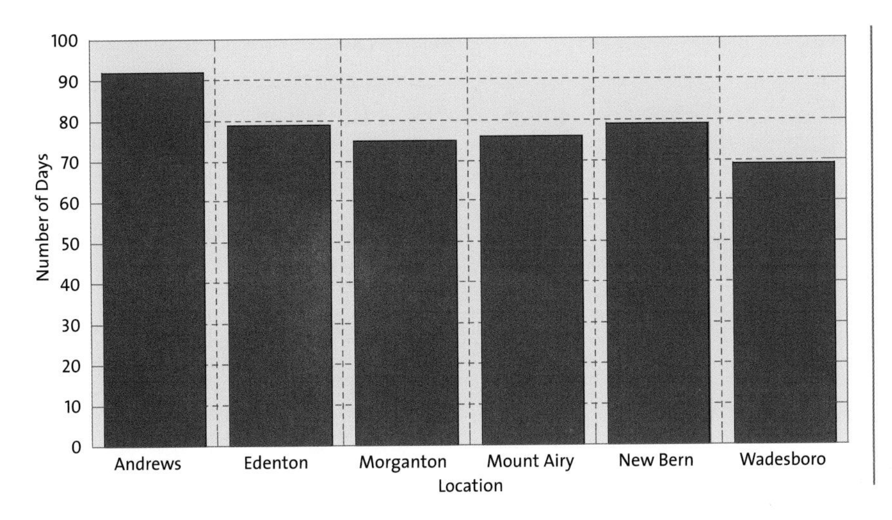

FIGURE 1.5. *Average number of days per year with more than 0.1" of precipitation for stations in a variety of North Carolina regions*

The multiplicity of elements and possible combinations presents a practical problem when we try to describe, understand, and forecast weather and climate. There are too many options, and we cannot possibly consider all the useful or interesting combinations or those that a particular individual may need for work or leisure activities. This is true for the weather forecasts we see and hear, and it is true for this book itself. In the book we discuss weather observations, analysis, and forecasting in broad terms, giving hints and examples that help you to take the information and tailor it to your own needs. To do this, we have adopted a dual strategy.

For the individual weather elements, we have adopted a climatological approach, emphasizing the values of different elements in various parts of the state. This involves using past observations to determine averages, extremes, or the probability that an event will occur. Figure 1.5 is an example of the kind of information that can be produced using this climatological approach. In the same way, later chapters contain, for example, a map of annual average temperatures across the state, a diagram showing how the temperature might vary with elevation in a mountain valley, and a table indicating the probability of having a frost after a particular date. We also give, in several places, hints on how you might use the climatological information for your own practical needs.

The alternative meteorological approach is used in this book to discuss the events that make up our daily weather. Here all the elements combine to create the sequence of events we know as the characteristic weather of a season or a region. So we shall discuss the sequence of weather commonly associated with the passage of weather fronts of various types, consider the meteorological character of hurricanes, and look at the atmospheric circumstances that lead to drought. Sometimes we shall give generalized, idealized descriptions; at other times we shall use actual examples. Both are needed, since they are complementary, and they will be woven together much of the time.

Understanding and Forecasting the Weather

Meteorology and climatology are sciences, and we increase our understanding of them, and make better forecasts, if we use a strategy commonly called the *scientific method*. In all science this primarily involves undertaking experiments, observing the results, and developing a theory to explain these results. New experiments are then devised and undertaken to test the theory. If the theory and the new results do not match, we must first check to ensure that the observations were taken correctly. If that turns out to be the case, we must change our theory. Theories that have stood the test of many experiments eventually become scientific laws and the basis for further theories and experiments.

Most scientists run their experiments in a laboratory. But it is extremely difficult to take the atmosphere, or even a small part of it, into a laboratory. In some specialized and limited circumstances, laboratory experiments are possible, and we have learned a lot about airflow around buildings and about urban air pollution from experiments in wind tunnels. However, most of the time this kind of experiment is not feasible. Rather, we must take observations of the actual atmosphere and conduct our experiment by making forecasts.

OBSERVING OUR WEATHER

These days we have a vast array of sensors available to observe the atmosphere. Satellites provide continuous global pictures and observations that, when suitably linked with direct observations from the land surface, can provide much of the information needed for weather forecasting. Radar systems and drifting buoys are increasingly being incorporated to provide additional data to support more detailed and more immediate forecasts. The surface observing stations themselves provide much of the quantitative information we need for these forecasts and almost all the information we have about climate and climate changes in the state. While we rarely see the instruments themselves, we are very familiar with their results: radar displays, satellite images, forecast maps, or tables and maps of climatic normals. In later chapters we will look at measurement methods for the different weather elements; Chapter 7 starts with a description of the whole observational system and how it is used to create forecasts. However, we must make a few comments about observations here, since they influence virtually everything that we can say about our atmosphere.

For much of the time we shall be using information from two networks of surface observations operated by the National Weather Service (NWS). These measurements constitute North Carolina's part of the official climate record of the nation. The first network, the *First Order Network*, observes a range of elements at hourly intervals but is confined mainly to airport sites. Only six stations (Asheville, Charlotte, Greensboro, Raleigh-Durham, Wilmington, and Cape Hatteras) have a period of continuous observation long enough for us to determine the climatic normals. So we shall use their information very frequently—not because they are typical or special, but because they are the locations for which we have information. The number of stations of this type has recently increased as instruments that record observations automatically have been developed. Some of these are maintained by the federal, state, or local governments, and others by the State Climate Office of North Carolina (see Fig. 1.6). They have only been operating a few years, but for some analyses we can use these newer stations.

The second network, the *Cooperative Network*, is much more extensive, involving around 200 stations. But it provides information about only daily maximum and minimum temperature and daily total precipitation. So we have a great amount of information about daily temperatures and precipitation, and we can readily draw maps of patterns across the state (see Fig. 1.7). Indeed, we have so much information that we often need to create simpler summaries. The state has been divided into eight homogeneous *climate divisions*, and we often use the averages of all stations in a division when discussing climatic conditions (see, for example, Fig. 2.7).

FORECASTS AND EXPERIMENTS

In the atmospheric sciences, we conduct an experiment to test our theories by making a forecast (see Fig. 1.8). The more we know about the atmosphere, the more accurate our forecasts should become. If we knew everything about the atmosphere, forecasts would be 100 percent correct 100 percent of the time. We are far from that situation. In fact, to get to that level we would have to measure every element everywhere on earth all the time, so there would be no room for us humans anyway. But meteorological research is progressing in much the same way as in the other sciences. Meteorology is different, however, because we participate

FIGURE 1.6. *An observing station maintained by the State Climate Office of North Carolina, measuring wind direction and speed (at the top of the tower), temperature and humidity (at the louvered cylinder on the arm to the left of the lower part of the tower), radiation (on the arm reaching toward the viewer), and precipitation (in the gauge on the separate pole to the left of the tower). The electronics are contained in the white box near the tower base, while the whole system is solar powered using the panel just above the electronics box.*

in, or at least hear about, the experiments every day. Most of the time I suspect we forget that the forecast is a great experiment, and we simply think of the practical aspects. But it is often useful to think of the forecast as an experiment, particularly when it goes badly wrong.

The forecasts are continuously improving. Twenty-five years ago we expected to get a forecast for today and tomorrow and a vague outlook for the following day. Now we expect reliable forecasts for four or five days ahead and outlooks for a further four or five days. These advances indicate that we have increased our understanding of the workings of the weather—of the science of the atmosphere. Further advance requires continued application of the scientific method, with more and better observations, more analyses of why forecasts worked or failed, and improvement in our theoretical understanding.

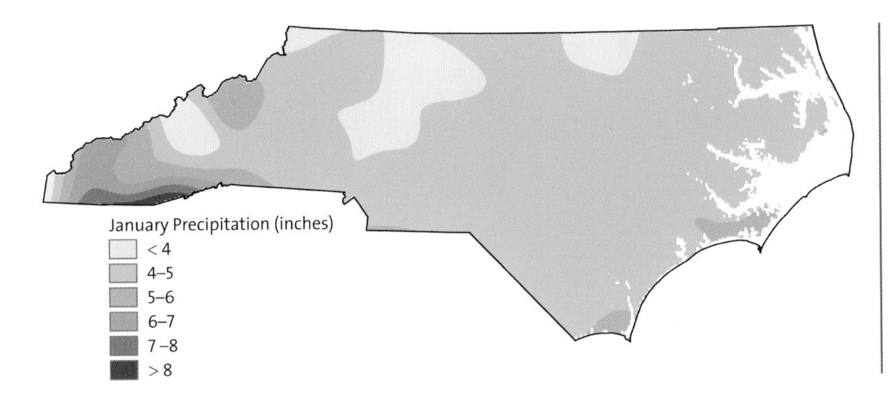

FIGURE 1.7. *Average total January precipitation across North Carolina, based on the observations of the National Weather Service Cooperative Observing Stations*

The theoretical part of meteorology nowadays almost always involves the use of atmospheric models. These are a set of mathematical equations that represent the physical processes creating and modifying the weather. They produce, among other things, maps of future weather that are the basis for the forecast. Like any model, they are simplifications of the real situation. We need to simplify not only because we do not have complete understanding or observations, but also because the computers that we use to solve the equations—to "run the models"—are not powerful enough to allow us to incorporate everything. We hope and trust that our simplifications incorporate the major processes while ignoring those that are of marginal significance. Each modeler, or modeling group, makes somewhat different choices for the simplifications; so there are various types of models, and they often give very different forecasts.

THE VARIOUS TYPES AND USES OF FORECASTS

For most people, forecasting is the main purpose of meteorology. There are probably as many uses for forecasts as there are people getting them. For practical purposes, therefore, we can break down the various types of forecasts available into a series of broad categories, based on how far ahead we look. In general, as the timescale increases, so does the space scale. Short-term forecasts may refer to a specific town or county, while the longest-term outlooks give estimates for the planet as a whole. In addition, the amount of detail available decreases as the timescale increases, and the methods we use to make the forecasts also change.

The shortest-range forecasts, in both time and space, are the mesoscale forecasts, often called nowcasts. These are developed from detailed information about current conditions, extend only a few minutes into the future, and indicate events in a specific community or even on a specific street. These forecasts depend on de-

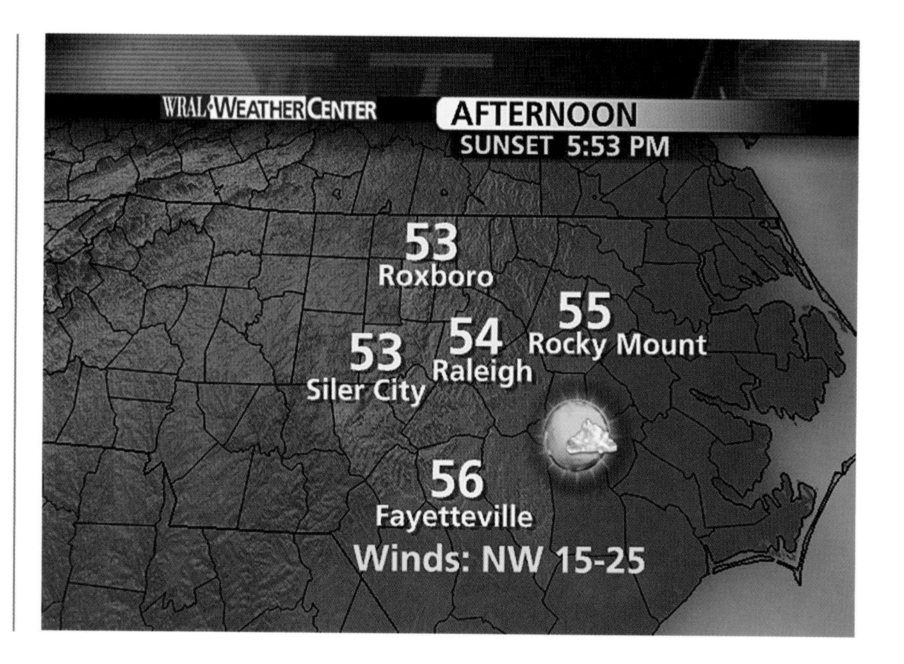

FIGURE 1.8. *A weather forecast, both an experiment testing all of our meteorological knowledge and a practical application of that knowledge (image courtesy of WRAL-TV)*

tailed computer models of atmospheric activity over a local area called mesoscale models. Probably the greatest benefit of these forecasts is that they give the few minutes warning needed to get out of the way of an approaching tornado.

The most familiar forecast type is the synoptic forecast, the daily weather forecast. Currently the area covered is roughly equivalent to several North Carolina counties, and the timescale allows a forecast for two to three days ahead, with an outlook for a few more days. We are slowly increasing our ability to give synoptic forecasts out to about ten days.

Although there are many uses for synoptic forecasts, a congressionally mandated role for the NWS is the provision of forecasts for the preservation of life and property. Fortunately, life-threatening weather events are not an everyday occurrence across our state, but they do occur. For such situations the NWS has developed the watch/warning system (see Table 1.1). Tornadoes, thunderstorms, flash floods, hurricanes, and heat waves all constitute threats where the provision of early warning can prevent the loss of life and minimize the loss of property.

For forecasts more than about ten days ahead, we have to take a different approach and compare the forecast conditions to the climatological normal. At present we can routinely provide information in this general way for temperature and precipitation for the upcoming months, giving the monthly and seasonal outlooks.

TABLE 1.1. Definitions of Watches and Warnings

Watch	Severe weather is possible within the designated watch area. Be alert.
Warning	Severe weather has been reported or is imminent. Take necessary precautions.

These refer entirely to the possibility that the month will have average temperature or total precipitation above, below, or near the climatological normal for the month. They say little directly about conditions within the month or season.

These long-range forecasts have only been available for a few years, and most of us do not use them very much in our everyday lives. They are becoming increasingly important to industry. For example, a utility company needs to generate more power in a cold winter than a warm one. The seasonal outlooks help the company to decide how much fuel to stockpile, aiding in the efficient use of its resources.

In order to look at weather far into the future, we must consider climate scenarios, a range of possible future climatic conditions, rather than individual, specific climate forecasts. The scenarios consider conditions on a continental scale and seasonal average conditions. Climate scenarios come from running atmospheric models very similar to those used in the development of synoptic or mesoscale forecasts. Like them, the scenarios contain numerous approximations and uncertainties associated with the science of the atmosphere, but unlike them, we cannot check every day to see whether we were right or not. Furthermore, changes in the character of the earth's surface and the composition of the atmosphere brought about by human activities add uncertainty. So it is unrealistic to suggest that we can accurately predict a single future climate, but we can talk about a range of possible climates within which we suspect the actual climate, when it finally arrives, will fall.

All of the above forecasts incorporate some mix of two kinds of forecast: *deterministic* and *probabilistic*. A deterministic forecast indicates exactly what, when, and where something is going to happen. Our aim is to give such a forecast whenever possible, because this gives the greatest amount of information. But for some atmospheric features we simply do not know enough or do not have good enough observations for this. Then we must use both the physical understanding we do have and knowledge gained from close analysis of past experiences with weather events to give the forecast. For example, we can say, "Nine times out of ten when we have seen this kind of weather pattern developing in the past, it has led to rain. So we have a 90 percent chance of rain this time." This is a probabilistic forecast. At the synoptic scale, temperature forecasts are usually deterministic, with the forecast indicating actual temperatures at various times in the future for various places in the forecast area. Precipitation, however, is usually given in terms of a certain percentage chance of rain—the term "chance," or something similar, always provides the clue that this part is probabilistic. In general, as we go to longer timescales, the proportion of the forecast that has to be probabilistic increases. Nowcasts are almost entirely deterministic; climate change scenarios are primarily probabilistic. Most forecasts contain a mixture of both.

It is very often possible to provide useful, highly detailed forecasts using the probability approach entirely. This method uses historical climatic data to provide what are often called "anytime" forecasts. As an example, Figure 1.5 shows that Edenton has about 80 days per year with more than 0.1" of precipitation. From that information alone we could suggest that the chance of Edenton having more than 0.1" of rain on any day is (80/365), or about 22 percent. Put another way, about 1 out of every 4.5 days has more than 0.1" of rain (climate statistics never seem to come out neatly). We could readily refine this by looking at data for a particular month and for days with any rain at all. We are well on our way to making our own forecast.

MAKING YOUR OWN FORECASTS

Why make your own forecasts, when the NWS and the TV stations—to say nothing of an increasing number of websites—provide them readily, routinely, and we hope, accurately? Let's change the question slightly and ask, "How often do we make our own weather forecasts?" The answer turns out to be "very often—probably every day." Some of our forecasts may be trivial—we rarely think of worrying about the chance of snow in a North Carolina summer—but some are almost certainly not. If it looks like rain and you need to arrive neat and tidy for that important meeting or interview, do you take an umbrella? The TV may give you lots of guidance, but in the end *you* have to make the forecast. Even more directly, we all plan weddings, plant gardens, or take vacations on an assumption—or a hope or a forecast—about the weather. Although only a few of us in the workaday world make forecasts or create weather information professionally, many of us may use forecasts directly or indirectly. People in the energy industry need forecasts for tomorrow and for the season ahead, and if a new plant is to be constructed, they even need estimates of climate change. Building contractors need information about rainfall rates to design adequate drainage systems, and agriculturalists need rainfall forecasts to help with irrigation scheduling. The list could go on and on, and some scenarios will be considered later.

So we all make our own forecasts, even though we probably take a great deal of guidance from TV, the NWS, a groundhog, or the farmers' almanac. Knowing more about the weather helps us to make better forecasts—or even to decide that our need for weather information is so important and complex that we need a professional to help us. This book aims to help us know more about the weather and to make better forecasts, or at the very least to make better use of forecasts made by others.

Producing Information about Weather and Climate

The production of forecasts and information about weather and climate requires the cooperation of many people and institutions. Amateurs and professionals have been working together in North Carolina for more than 150 years to produce our present knowledge and to create the system for generating and disseminating the weather forecasts and information we now have available.

THE NORTH CAROLINA WEATHER SERVICE: FORECASTS BY MAIL AND RAIL

The earliest observations in North Carolina for which we have well-documented records are from Chapel Hill in 1820. This station, with minor changes in location, continues to this day. Other stations started in the 1830s and 1840s. In those early days, observers were usually university or school educators, physicians, or clerics observing primarily so that they could determine the nature of our national weather. They were working in the tradition of Thomas Jefferson and Benjamin Franklin, both of whom sought to understand the weather so that agriculture and commerce could be improved.

As the nation expanded westward and it became obvious that the climate was not the same everywhere, the Army Signal Service began to take observations and transmit them via telegraph to a "weather bureau" in Washington, D.C. Here the data were analyzed to provide the first climatological descriptions of the United States. These analyses also began to hint at the presence, nature, and movement of storms. These ideas led to the beginnings of forecasting.

In North Carolina, six Signal Service observing stations were established in the early 1880s. Four of the stations were on the coast to give early warning of the arrival of storms, particularly hurricanes. The other two, at Raleigh and Charlotte, were located close to population centers to monitor inland storm movement. About the same time, the North Carolina Agricultural Experiment Station, established in 1877, began to collect, archive, and analyze the observations made by volunteers around the state. In 1886 the federal and state activities were combined, and the North Carolina Weather Service was born. A decade later this service became part of the U.S. Weather Bureau within the federal Department of Agriculture. The Weather Bureau continued and expanded the dual role of providing both daily weather forecasts and longer-term term climatological analyses. Indeed, from the beginning the two different types of observing networks required—first order and cooperative—have been used. But the techniques involved have changed dramatically.

Weather forecasting in the days of the North Carolina Weather Service—the last two decades of the nineteenth century—depended on the telegraph. A network of observers along the telegraph lines was established. Since the telegraph was closely associated with the railroads, the distribution of observers was far from even, but it did reflect the population distribution of the state. Each afternoon the observers telegraphed their observations to the Weather Service headquarters in Raleigh. In the early years only a few out-of-state observations were incorporated, but gradually the area involved increased. Forecasting primarily meant mapping the pressure observations, tracking the low pressure areas as they moved as "storms" across the state, and projecting their tracks for the next twenty-four hours. As knowledge increased, temperature and precipitation forecasting was incorporated, and it became possible to forecast the occurrence of "cold waves" in spring and fall, vital information for a state that was still largely agricultural. The forecast was usually made in the early evening and was valid for the twenty-four-hour period commencing in the morning of the following day.

Once created, the forecast had to be disseminated. By 1893 two methods were in place. The primary method used was the telegraph. Each evening the forecast was filed at the Raleigh telegraph office so that it would be available to all recipients by 8:00 A.M. the following morning. The recipients, mainly signals officers at the main railroad depots who acted for the Weather Service in a voluntary capacity, then disseminated the forecasts locally. In most cases the forecasts were posted on the public notice board commonly associated with the depot. Where possible, a signal flag was also hoisted (see Fig. 1.9), and sometimes an audible storm warning signal was also given. The second dissemination method was used mainly for the smaller, intermediate communities along the railroad lines. Either the Raleigh office or one of the telegraph forecast recipients created postcards containing the forecast and mailed them in sufficient time to catch the early trains. The local postmaster would then display the forecast and, in some cases, hoist signal flags.

MODERN WEATHER SERVICES

By 1900 the framework of a national meteorological system was well established. Since then the basic structure of services has continued with only minor administrative changes. But there has been a major expansion in our ability to observe and understand the atmosphere and in the range of services offered.

For much of the early part of the twentieth century the Weather Bureau continued as part of the U.S. Department of Agriculture, with most of the research and forecasting efforts going toward linking weather and crop production. However, probably the most visible advance in this period was the appearance of weather fronts on weather charts. Although we take them for granted and expect them on most weather maps, and although they help explain many weather features, even now we are still a long way from explaining how they work.

The increase in aviation, particularly immediately after the Second World War, meant that the aviation industry was a new client for weather services, while aircraft weather observations began to give a three-dimensional picture of the atmosphere. Meteorological knowledge expanded rapidly. A further boost occurred when satellite information became available on a routine basis. More accurate and longer-range forecasts began to appear. These improving forecasts have practical applications in a wide range of activities, not only in the traditional areas of agriculture and aviation, but also in the transportation, water resources, health, construction, and energy sectors of our economy. And they have an impact in our own everyday lives.

The most recent advances in meteorology have been spurred by developments in electronics and computer technology. New electronic instrumentation is increasing both the number of stations involved in the surface observing networks and the frequency and types of observations made. Modern communications technology allows rapid transmission of these observations to the national forecasting center and helps to increase forecast accuracy. Meanwhile, the increase in the power of computers themselves is allowing the development of ever more detailed and sophisticated mathematical models of the atmospheric processes driving our weather. At the same time, detailed observation and analysis by humans is still a vital part of the system. The result is a forecast network with many branches. Some give very general forecasts for everyone in a wide area, some provide highly de-

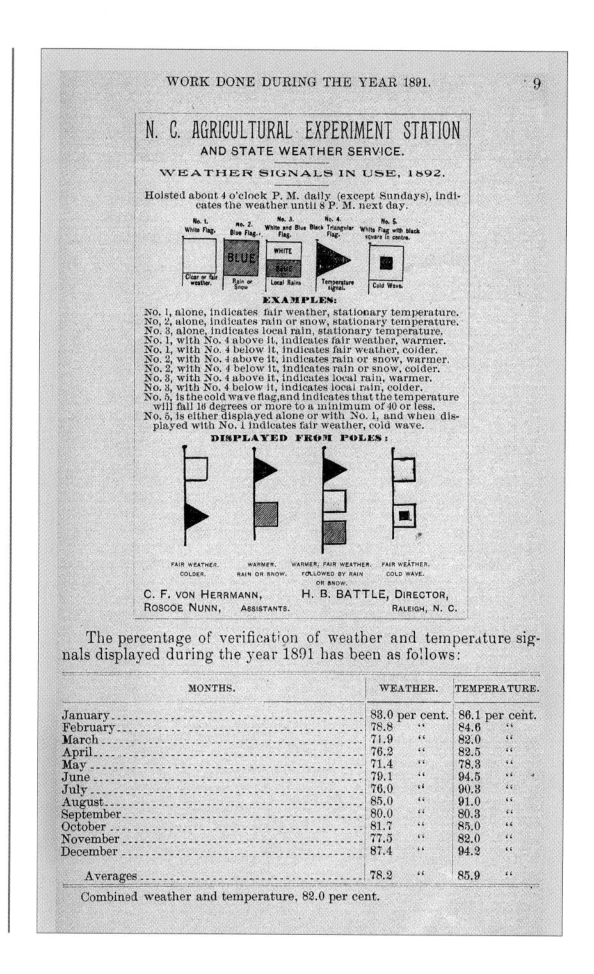

WORK DONE DURING THE YEAR 1891. 9

N. C. AGRICULTURAL EXPERIMENT STATION
AND STATE WEATHER SERVICE.

WEATHER SIGNALS IN USE, 1892.

Hoisted about 4 o'clock P. M. daily (except Sundays), indicates the weather until 8 P. M. next day.

No. 1. White Flag. — Clear or fair weather.
No. 2. Blue Flag. — BLUE — Rain or Snow
No. 3. White and Blue Flag. — WHITE / BLUE — Local Rains
No. 4. Black Triangular Flag. — Temperature signal.
No. 5. White Flag with black square in centre. — Cold Wave.

EXAMPLES:

No. 1, alone, indicates fair weather, stationary temperature.
No, 2, alone, indicates rain or snow, stationary temperature.
No. 3, alone, indicates local rain, stationary temperature.
No. 1, with No. 4 above it, indicates fair weather, warmer.
No. 1, with No. 4 below it, indicates fair weather, colder.
No. 2, with No. 4 above it, indicates rain or snow, warmer.
No. 2, with No. 4 below it, indicates rain or snow, colder.
No. 3, with No. 4 above it, indicates local rain, warmer.
No. 3, with No. 4 below it, indicates local rain, colder.
No. 5, is the cold wave flag, and indicates that the temperature will fall 16 degrees or more to a minimum of 40 or less.
No. 5, is either displayed alone or with No. 1, and when displayed with No. 1 indicates fair weather, cold wave.

DISPLAYED FROM POLES:

FAIR WEATHER. COLDER.
WARMER. RAIN OR SNOW.
WARMER, FAIR WEATHER. FOLLOWED BY RAIN OR SNOW.
FAIR WEATHER. COLD WAVE.

C. F. VON HERRMANN,
ROSCOE NUNN, ASSISTANTS.
H. B. BATTLE, DIRECTOR,
RALEIGH, N. C.

The percentage of verification of weather and temperature signals displayed during the year 1891 has been as follows:

MONTHS.	WEATHER.	TEMPERATURE.
January	83.0 per cent.	86.1 per cent.
February	78.8 "	84.6 "
March	71.9 "	82.0 "
April	76.2 "	82.5 "
May	71.4 "	78.3 "
June	79.1 "	94.5 "
July	76.0 "	90.3 "
August	85.0 "	91.0 "
September	80.0 "	80.3 "
October	81.7 "	85.0 "
November	77.5 "	82.0 "
December	87.4 "	94.2 "
Averages	78.2 "	85.9 "

Combined weather and temperature, 82.0 per cent.

FIGURE 1.9. *Weather forecast signals used by the North Carolina Weather Service toward the end of the nineteenth century (North Carolina Agricultural Experiment Station,* Climatology of North Carolina from Records of 1820 to 1892 *[Raleigh, 1892])*

tailed forecasts for people with particular information needs or in a specific area, some predict events only a few minutes in the future, and some look decades—or even centuries—ahead.

One wonderful consequence of these recent technological advances is that a vast amount of weather information is easily and immediately available to anyone with access to the World Wide Web. Appendix A gives a list of the current major information sources. There is a mix of official government information, university and commercial weather analyses, educational materials, and media resources. We have already used some of these earlier in this chapter, and they will appear in all later chapters as we describe and seek to explain the weather and climate across our state. We all have access to virtually the same information as do the forecasters at the local NWS office or at your favorite TV station. That might not mean we can make forecasts as well as they can, but it is always fun to try.

chapter 2

Sunshine, Seasons, and Temperature

Sunshine and temperature are major features of our weather. They are also closely related to each other because the amount of sunshine largely controls our temperatures. All we have to do is stand in the sun or recall that the long days of a North Carolina summer are hot to experience the relationship. In this chapter we explore that relationship, looking first at sunshine and how it affects what we see, then at how it influences temperature, and finally at our North Carolina temperatures.

Sunshine in North Carolina

THE AMOUNTS WE RECEIVE

From our own experience, we expect the amount of sunshine we receive at any place to depend on the length of daylight and the amount of cloud coverage. This is indeed the case. If we look at the total amount of sunshine we receive each month, the change in day length clearly imposes a seasonal pattern (see Fig. 2.1a). Most places in North Carolina have about 275 hours of sunshine in each summer month but around 175 hours per month in winter. During winter and spring, Wilmington is sunniest, while in summer and fall, bragging rights go to Charlotte. Throughout the year, Cape Hatteras tends to have less sunshine than most other parts of the state.

While the total amount of sunshine in a month is most important when we are connecting sunshine and temperatures, for most of us the real question is “Is the sun shining?” The best answer comes by looking at the percentage of possible sunshine each month (see Fig. 2.1b). This figure indicates the amount of sunshine as a fraction of the total that would occur on a completely cloudless day in a given month. It removes the effect of day length and emphasizes the relationship

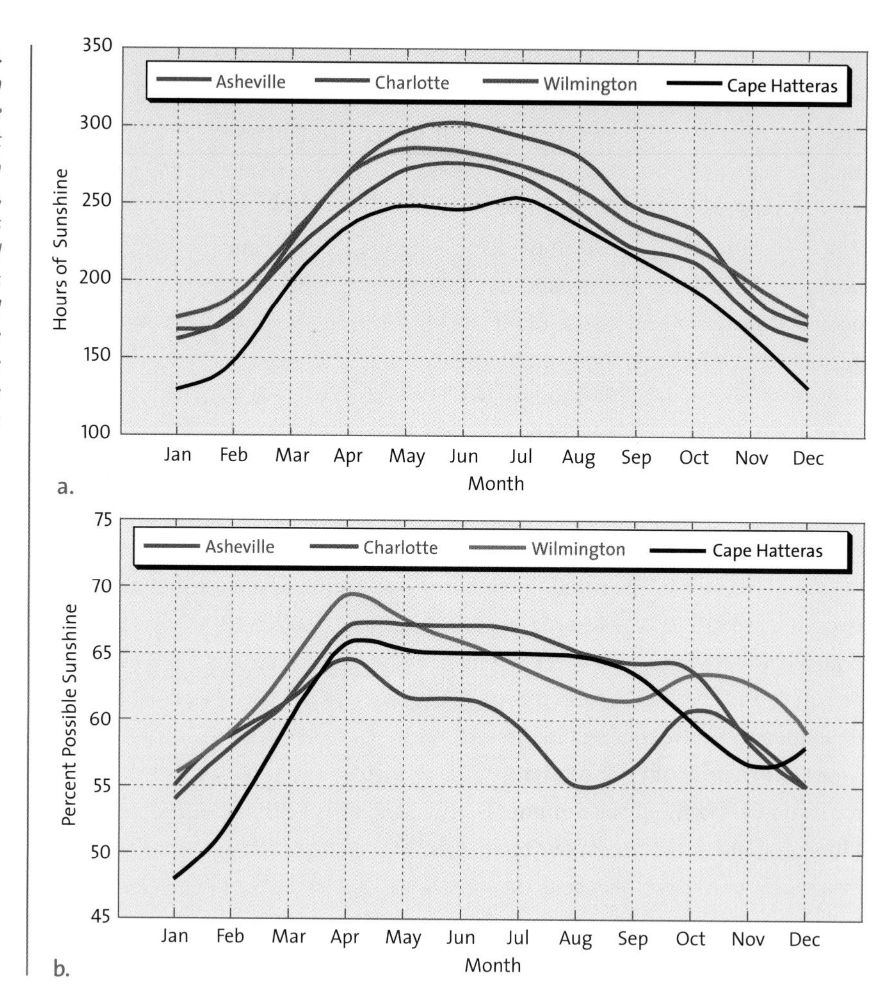

FIGURE 2.1. *Long-term monthly average sunshine amount for selected North Carolina locations, expressed as (a) the total number of hours per month and (b) the percentage of possible sunshine hours*

between clouds and sunshine. The relative positions between stations do not change much, but the diagram shows that for most of the state the sun is shining more than 55 percent of the time, but rather less in winter than in summer. One of the main things this emphasizes is that the seasonal change in day length controls our temperatures.

WHAT CAUSES THE SEASONS?

The cause of seasonal changes in the length of a day is the astronomical relationship between the sun and the earth. The earth revolves around the sun, making a complete circuit once a year. The earth also rotates on its own axis, making one rotation per day. This rotation axis is a line that joins the North and South poles and passes through the center of the earth. The axis is tilted so that in the Northern Hemisphere the North Pole is inclined toward the sun during the summer. The result for us in North Carolina is that for more than half the year we are on the sunlight side of the earth and have long days. In winter the North Pole is inclined away from the sun, and our nights are long.

On the surface of the earth we see these events as a cyclic change in both the position of the sun in the sky and the length of daylight. The extreme summer tilt occurs at the *Summer Solstice* (June 21–22). Prior to this date, each day has been getting longer and the noon sun has been slightly higher in the sky. After it, days become shorter and the sun sinks lower in the sky. On the solstice (a term derived from Latin *sol*, the sun, and *sistere*, to cause to stand still), the noon sun reaches its highest position in the sky and seems to stand still. That day is also the longest of the year. The actual length depends on latitude; the farther north, the longer the day (see Table 2.1). At the opposite extreme, the *Winter Solstice* (December 21–22) is the shortest day, and the noon sun is at its lowest point for the year. The two intermediate positions are the Vernal Equinox (March 20–21) and Autumnal Equinox (September 22–23). (The term "equinox" is derived from Latin *aequus*, or equal, and *nox*, night.) On this date, every place on earth has twelve hours of daylight and twelve hours of night.

Astronomically the seasons start at a particular equinox or solstice. Summer, for example, begins at the Summer Solstice and lasts until the Autumnal Equinox (June 21–September 22). If our temperatures in North Carolina followed the sun exactly, they would have a peak on June 21 and a minimum on December 21. However, factors other than sunlight affect temperatures, and there is a lag between the absorption of sunlight and the rise in temperature. So maximum temperatures usually occur in July or, in some years, even in August. The lowest temperatures occur in January or February. Because of the lag and variability, it has become standard to define the *meteorological* seasons in the following way:

spring: March, April, May
summer: June, July, August
fall: September, October, November
winter: December, January, February.

SUNLIGHT, VISIBILITY, AND COLOR

All the colors that we see, whether the "Carolina blue" of the sky, the familiar white of clouds and green of grass, the spectacular red of a sunset, or the faint tinge that

TABLE 2.1. Lengths of the Longest and Shortest Days at Three Latitudes within North Carolina

LATITUDE (°N)	LONGEST DAY (HOURS:MINUTES)	SHORTEST DAY (HOURS:MINUTES)
36 [a]	14:27	9:33
35 [b]	14:21	9:39
34 [c]	14:16	9:44

[a] close to the Triangle and Triad cities
[b] just south of Charlotte and New Bern
[c] just south of Cape Fear

gives our Blue Ridge its name, come from the interaction between sunlight and the atmosphere or the earth's surface. Sunlight is composed of a series of infinitesimally small particles called photons. These travel in an oscillating (wavelike) fashion, vibrate very rapidly over a very small distance, and are much too fast and too small for us to see. The photons can have a range of oscillation rates, and our bodies respond in various ways to particular rates. Many rates are detected by the eye and seen as colors. There is a particular oscillation rate for every color of the spectrum. Together they make up the visible spectrum or, more simply, sunlight. There are also photons with faster rates, the ultraviolet, which if we are not careful, we experience as sunburn. The photons that vibrate more slowly than red light are called infrared, and our skin converts their energy rather readily into heat. The energy in the full spectrum we call the *solar radiation*.

The various colors and hues we see at any time depend on the mix of photons entering our eye. The sun emits the full spectrum and produces white light. Satellite and astronaut pictures from above the atmosphere clearly show this white sun. However, as the photons pass through the atmosphere toward us, the photon mix changes, and colors begin to appear. Atmospheric obstacles to the photons, whether they are air molecules, cloud drops, or pollution particles, can reflect, absorb, transmit, or scatter photons. The amount of each type of interaction, and the colors involved, depends on the type of obstacle.

Ordinary air molecules are just the right size to scatter the photons we see as blue light. Some of these scattered photons arrive at our eyes. They come from all directions, and we see a blue sky. The direct beam of the sun has lost its blue photons and usually appears yellow. Near the beginning and end of the day, when the sunlight has a longer path through the atmosphere, there is more chance for scattering of photons other than blue. The result is the reddish color of the setting sun.

The creation of the blue of the Blue Ridge is another consequence of this scattering of light by air molecules. The greater the distance an object is from our eyes, the bluer the image. Eventually the scattering leads to a blurring of contrasts between adjacent images so that we can no longer distinguish between them. This creates the limit of visibility. In deserts this limit is about 90 to 100 miles. But in North Carolina the trees of our wooded landscape emit extra molecules that enhance the atmospheric scattering and increase the blue appearance. They make the limits of visibility about 70 miles. This effect is entirely natural; the Blue Ridge was known by that name long before industrialization raised any concerns about pollution.

During the last fifty years the annual average visibility has varied considerably. All three stations in Figure 2.2 show a decreasing trend from the early 1950s to the mid-1970s. Then, possibly as the result of various Clean Air Acts, visibility increased, markedly in Asheville's case, only to decrease again. More recently, Charlotte has seen the clearest air for many years, but there has not been a comparable increase in visibility at the other stations. So there is no obvious trend. The averages are, of course, all well below the seventy-mile limit of visibility for the area. Probably that has always been the case. Unfortunately, we do not have any reliable long-term observations of visibility from a western mountaintop.

In contrast to air molecules, water droplets or ice crystals in clouds reflect almost all colors. A thick horizontal cloud appears to be very bright white from above, but seen from below it transmits few photons of any color and so looks black or gray. A vertical cloud wall may look very bright when we are in a position to see the reflected sunlight, or it might appear very black and threatening when it is between us and the sun.

Few natural surfaces on earth reflect all photons equally. The photons that are reflected most determine the color of a surface. Grass, for example, reflects only those photons we sense as green light. It transmits a few of the others and absorbs the rest, using the absorbed energy for photosynthesis and heating.

Most objects reflect or absorb sunlight in a very thin layer right at their surface. Water is different. There is some reflection at the surface, along with some absorption and transmission. The portion that is transmitted then interacts with the layer of water just below the surface, which again reflects, absorbs, and transmits. This layer-by-layer process continues until all the photons have been absorbed or reflected. This final depth depends on how much suspended and dissolved material there is in the water. The suspended material is primarily responsible for the reflection and so dictates the water color we see. Many Tar Heel rivers look tan or brown because of the soil they are carrying to the sea. Water

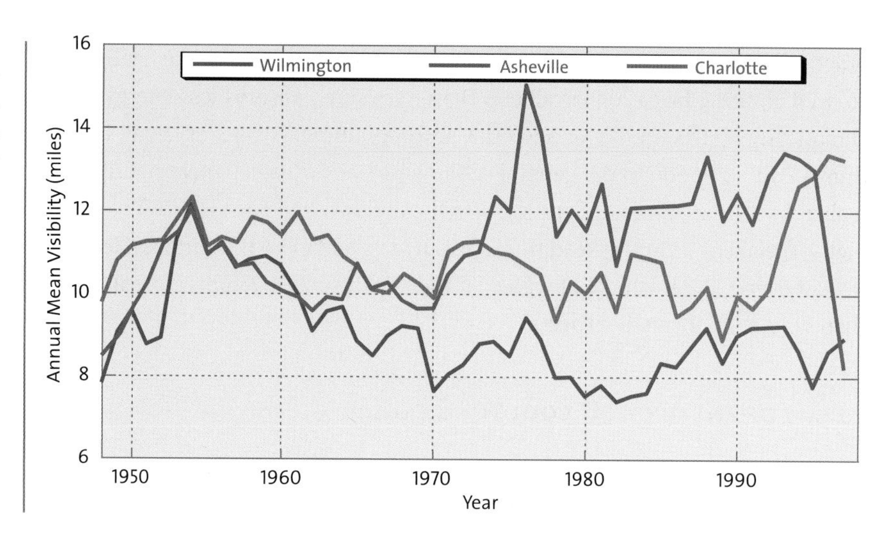

FIGURE 2.2. *Trends in annual average visibility since 1948 for three selected stations in North Carolina*

bodies with little suspended material tend to mimic the color of the sky and appear bluish.

There is a further twist that makes water unique. The amount of reflection depends on the angle of incidence of the photon. Photons that arrive more or less vertically are most readily absorbed. Most of those coming in at a low angle are reflected. As a result, if a day is calm and sunny, at dawn and dusk a body of water will have a mirrorlike surface, and we can see the reflected sun. At any time when it is sunny but not calm, certain facets of the choppy waves will reflect the direct sunlight, while others will absorb photons. As the waves change, so will the reflection, and a sparkling water surface will result.

The Cause of Temperature: Linking Energy and Temperature

SURFACE HEATING BY SOLAR RADIATION ABSORPTION

Photon absorption causes heating. However, it is much easier to measure reflection than absorption, and we always deal with reflection. The percentage of the incoming radiation that is reflected is called the *albedo*. Surfaces with high albedo absorb little solar radiation and tend to stay cool; those with low albedo values commonly reach higher temperatures. Water has a low albedo, most rural and urban surfaces are somewhere in the middle, and ice and snow have a high albedo. Snow and ice, especially newly fallen snow, reflect virtually all photons. With no absorption, snow does not heat—and melt—in sunlight. As far as climate is concerned, this is a major reason why the poles remain cold even in the summer. Perhaps more practically, it is why we are able to ski on a sunny day. Equally practically, light-colored clothing has a higher albedo than dark suits, absorbs less energy, and is climatically more suitable for a North Carolina summer. Water, as usual, is something of an exception to the general rule relating absorption to heating. Although it absorbs a great amount of energy—except at dawn and dusk when the albedo is high—the absorption is spread through a deep layer below the surface. Heating is spread over a large volume of water, so the rise in surface temperature is smaller than we would normally expect.

FORMS OF ENERGY THAT COOL THE SURFACE

The absorption of photons leads to heating; but this cannot be the only factor that determines surface temperature, or our land would get hotter every day. There must be other processes acting to prevent continuous heating. There are four such processes that, like solar radiation, are streams of energy with specific characteristics (see Fig. 2.3). The ground heat flow moves energy to and from the deeper layers of the ground or ocean, with the flow going from the hotter areas to the cooler ones. This is the flow that warms the soil in spring and may lead to frozen ground in winter. The sensible heat flow—the heat you can sense or feel—also travels from hot to cool areas, mainly from the warm surface into the cooler atmosphere. The latent flow is the heat hidden or latent in water vapor so that evaporation takes energy from the surface of the earth and releases it during condensation as clouds. Finally, the terrestrial radiation behaves rather like sunlight, but we cannot see it with our eyes. We normally sense this energy when our body converts it to heat. The long-term balance between these streams maintains our planet at a relatively constant temperature but with variations over time and place. Since the solar radiation tends to be the main driving force and the other flows are a response to it, the seasonal pattern of temperature bears a strong resemblance to the seasonal pattern of sunlight and day length. Similarly, temperature changes within a day follow the course of the sun through the sky. The other flows, however, ensure that there are many modifications of these basic patterns.

THE GREENHOUSE EFFECT AND NORTH CAROLINA'S CLIMATE

One of the energy forms depicted in Figure 2.3, *terrestrial radiation*, is very similar to solar radiation and sunlight. In fact the only difference is that the photons associated with terrestrial radiation vibrate at a much slower rate—too slowly for our eyes to sense. All objects on earth—ground, sky, trees, buildings, and our

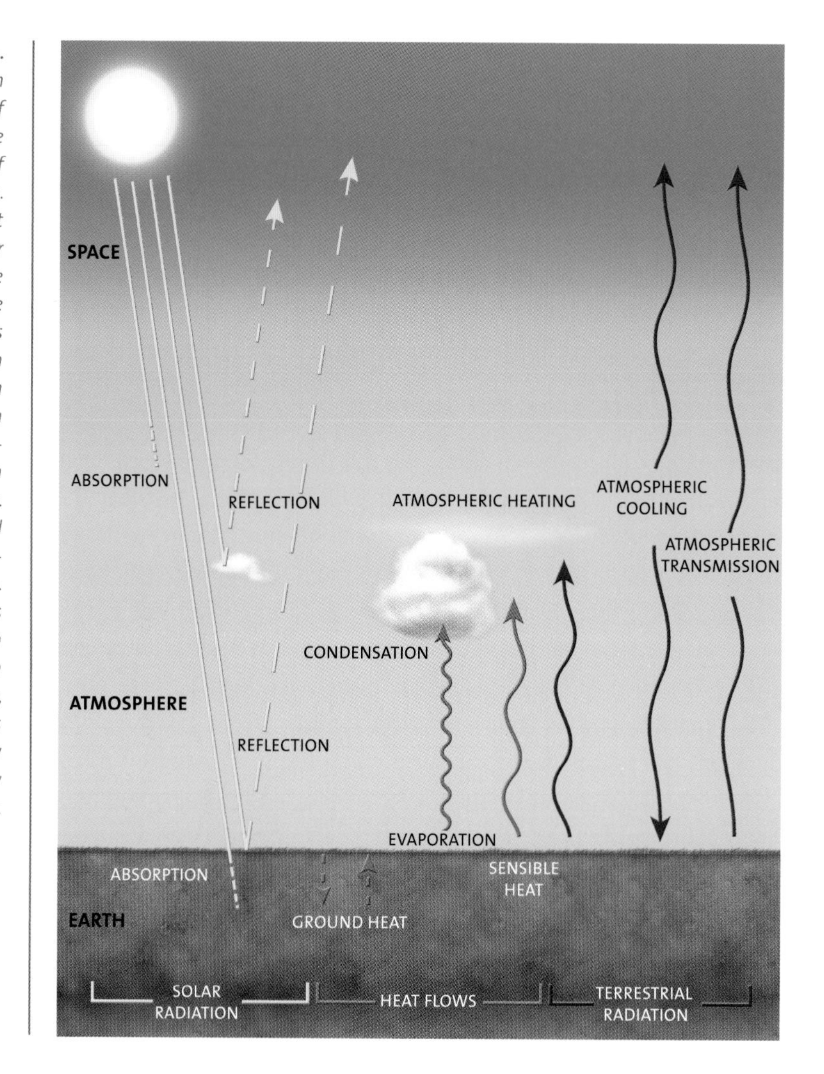

FIGURE 2.3. *Schematic diagram of the flows of energy that create the weather of planet earth. Reading from left to right, solar radiation from the sun warms the earth. Energy is transferred, in several ways, from the warm earth into the atmosphere, which in turn warms. Energy is returned to space as terrestrial radiation. The actual amounts involved vary with time and from place to place, and the winds move energy horizontally as a result.*

bodies—emit this terrestrial radiation. In addition, all objects also absorb this radiation. The absorption leads to warming; the emission, to cooling.

Solar and terrestrial radiation interact with the atmosphere to produce the *greenhouse effect* (see Fig. 2.4). Of the solar radiation that arrives at the top of the atmosphere, only a small portion is absorbed as it passes through the atmosphere. Most reaches the earth's surface. Some of this is reflected back to space, and the remainder is absorbed. This provides heating for the surface, which causes the temperature to rise. The hotter the surface, the more terrestrial radiation it emits. This radiant energy is sent upward into the atmosphere, where atmospheric gases absorb a large fraction of it. Absorption again leads to heating and an increase in

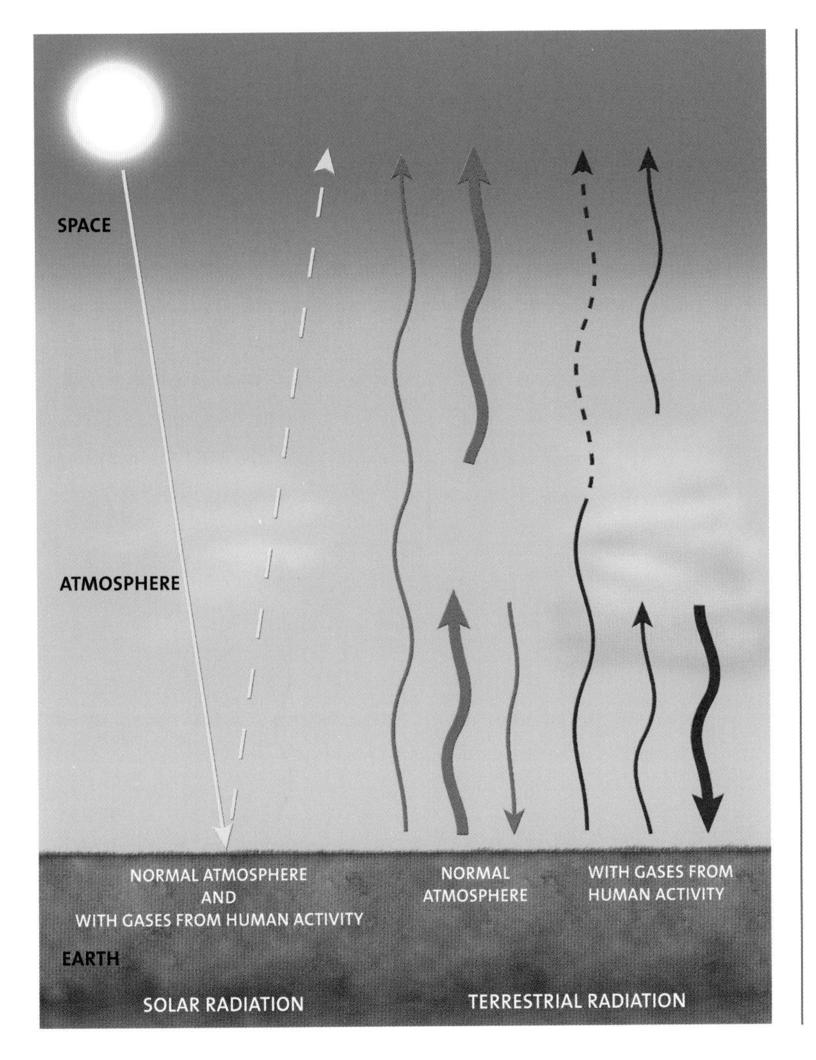

FIGURE 2.4. *Schematic diagram of the greenhouse effect. Solar radiation passes through the atmosphere without much absorption, while much of the terrestrial radiation is readily absorbed and emitted by greenhouse gases in the atmosphere. The greater the concentration of the greenhouse gases, the higher the temperature.*

emission. Some of this emitted energy passes up through the atmosphere and out to space, while the rest is returned to the earth's surface, where it is absorbed. This increases the surface heating and encourages the emission of more radiation. So a continuous cycle of terrestrial radiation exchange is set up. The continuous loss of terrestrial radiation to space promotes a cooling of the planet. This loss is compensated by the gain of absorbed solar radiation. So a rough balance is achieved, and the planet has, on a long-term basis, a reasonably constant temperature. That temperature depends on the amount of greenhouse gas in the atmosphere; the greater the concentration, the higher the temperature. With the current concentration, the average temperature of our planet and, partly by coincidence, of our state as well is about 60°F. If there were no greenhouse gases, those average temperatures would both be close to 5°F, and life would be much more difficult, if we could support it here at all.

The greenhouse effect is entirely natural, and the main gases involved—water vapor, carbon dioxide, and methane—have been constituents of the atmosphere for eons. However, with increased human activity—agricultural and industrial—the concentrations of both carbon dioxide and methane are increasing. In addition, some "new" greenhouse gases, such as the chlorofluorocarbons used in air conditioners, have been introduced. At the same time, global temperatures have been rising, and the two facts have been linked. At present the amount, rate, and location of warming resulting from an increased greenhouse effect is not fully clear. North Carolina—and the southeastern United States in general—is, in fact, one of the few places on earth to exhibit no clear warming trend over the last century.

FORECASTING NIGHTTIME TEMPERATURES

Although clouds are not "greenhouse gases," at night they do act in much the same way in influencing our temperatures. On a cloudy night the clouds absorb almost all of the terrestrial radiation emitted by the earth. They reradiate much of it back to the surface, so that the net loss of energy is small. Consequently, the overnight temperature fall is also small. Cloudy nights tend to be warm nights. On a clear night, however, relatively little of the outgoing terrestrial radiation is absorbed by the atmosphere and returned to earth. There is, therefore, a loss of energy from the surface, which cools rapidly. You can readily use this information to make an overnight temperature forecast. A clear evening sky suggests a cold night; evening clouds indicate warmth. Of course, if a cloud deck drifts across your clear evening sky overnight—or if an evening cloud layer disappears overnight—your forecast will go awry.

HOW SURFACE TYPES AFFECT TEMPERATURE

The pattern of temperature across the day and across the year at any point depends on the nature of the surface at that place. If we have a dry surface—whether a desert or a parking lot—there can be no evaporation. Consequently there can be no latent heat flow, and one of our four methods of removing absorbed energy is no longer available. The other flows have to work harder, with the result that the surface temperature gets higher.

In the broadest terms there are two basic surface types for our planet: land and water. Water, being able to exchange energy by both latent and sensible heat, warms and cools much more slowly than land. This effect, called the *continentality effect*, means that the interiors of continents are very hot in summer and very cold in winter. The oceanic margins have warm summers and cool winters. All of our state is on an oceanic margin, but as we go inland toward the Mississippi Valley, we get into an increasingly continental temperature regime.

The differences between land and water surfaces can also influence weather on a smaller scale. In our state we have three general surface types: the wet and swampy areas near the coast, which tend to have small daily and seasonal temperature swings; the dry and sandy region of the Sandhills, where the changes are much more marked; and the rest of the state, which has intermediate surfaces and temperatures. These become important when we look at the weather of our state on a regional scale.

North Carolina's Temperatures

OUR LINKS TO GLOBAL TEMPERATURE PATTERNS

The basic pattern of global temperatures is determined by the distribution of solar radiation, the effects of continentality, and altitude (see Fig. 2.5). Daily total solar radiation depends on both the length of the day and the height of the sun in the sky, both of which vary with season and latitude. In summer, day length increases as we go toward the North Pole, while the solar intensity is at a maximum around the Tropic of Cancer (23.5°N, between Florida and Cuba). Combining these two factors, the basic temperature distribution during summer in the Northern Hemisphere shows a broad maximum in the northern tropics and a gradual decrease toward the relatively warm North Pole. At the same time, the Southern Hemisphere is experiencing winter, and there is a rapid poleward decrease in temperatures and a very cold pole. The pattern is reversed as we go from summer to winter.

North Carolina, with its midlatitude location, is influenced by the continentality effect. Our predominant wind direction is from west to east, so for much of the time the air arriving over us has traveled a long distance over the North American continent. This incoming air is generally cold in winter. As it continues past us on a journey toward western Europe or North Africa, it blows over a warmer ocean. Thus a warm airstream arrives at the west coast of those continents. As a general result, in the area of the westerlies, the west coasts of continents tend to be warmer in winter than the east coasts at the same latitudes. In summer the continentality

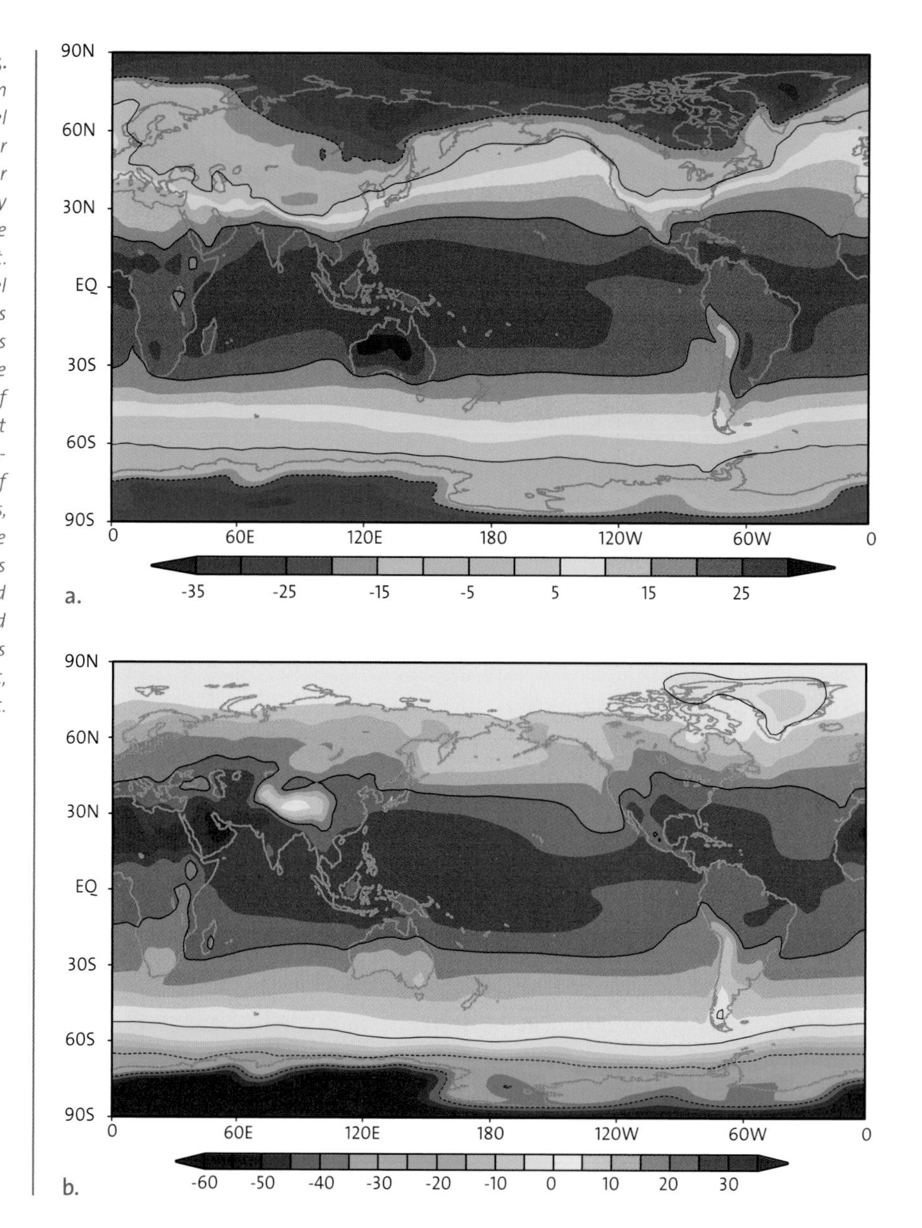

FIGURE 2.5. *Global distribution of mean sea level temperatures for (a) December through February and (b) June through August. Using sea level values emphasizes the factors controlling the distribution of temperature but can give a misleading picture of actual temperatures, particularly for the high-altitude areas of Tibet, Africa, and parts of the United States. The lines indicate -20°c, 0°c, and +20°c.*

effect works in the opposite way, and for similar latitudes, west coasts are cooler than east coasts. Put another way, this also means that our winter temperatures are similar to those of more northern latitudes on west coasts, and our summers are more akin to places much closer to the equator.

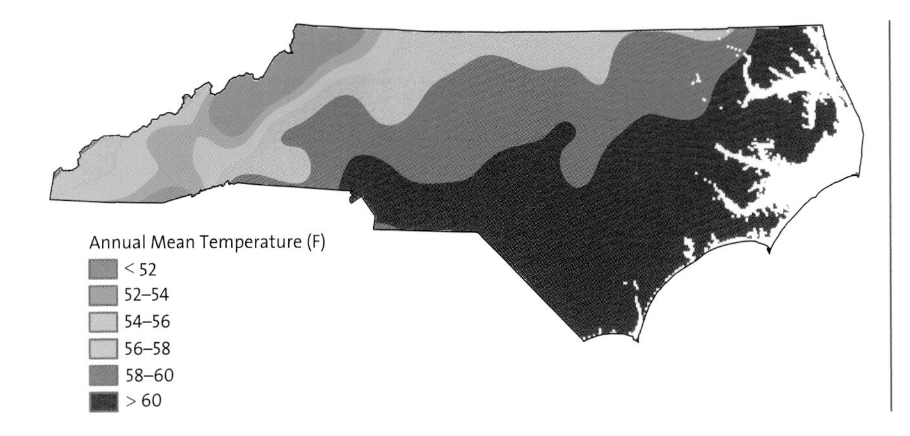

FIGURE 2.6. *Distribution of long-term average annual mean temperature across North Carolina. This map, along with most of the others in this section, is based on thirty years of daily temperature observations at about 200 stations around the state.*

AVERAGE TEMPERATURES AROUND THE STATE

For North Carolina as a whole, we can consider a variety of temperature characteristics. Figure 2.6 shows the annual mean temperature across the state. This map tells us many things, and some may not be very surprising. That temperatures decrease from east to west—that it is colder in the mountains—is probably no surprise. Similarly, we would probably expect the north to be somewhat colder than the south. But the pattern is not even. The Sandhills area west of Fayetteville, for example, seems to continue the coastal warmth some distance inland, possibly because of the dryness of this sandy region. The Outer Banks are rather cool, probably because of the cooling effects of the oceanic waters. Further, although it is difficult to be definite with a map of the whole state, temperatures appear to change rapidly with location in the west, but only slowly in the east. To provide a very general summary, mean annual temperature ranges from the low 50s in the northern mountains to the low 60s along the southern Coastal Plain.

The monthly and seasonal patterns of average temperature across the state are similar to the annual pattern. So instead of a series of almost identical maps, we can refer to the monthly average temperatures of each climate division (see Fig. 2.7). All areas have their highest temperatures in July and their lowest in January, as we would expect from experience as well as from the earlier consideration of energy exchanges. The mountains stand out as the cold area throughout the year. In winter the Piedmont divisions are generally cooler than those of the Coastal Plain, but in summer the differences between them are small.

The average daily minimum temperatures in January (Fig. 2.8a) range from the low 20s in the mountains to the low 30s near the coast and approach the

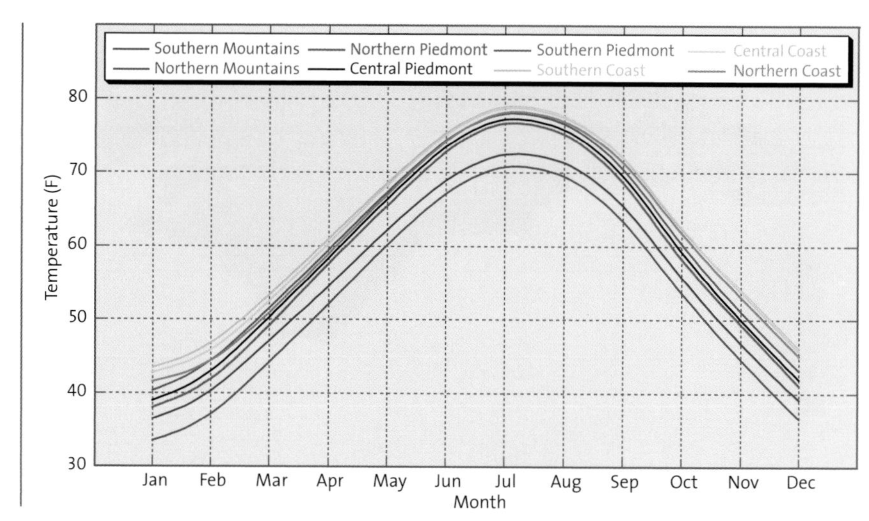

FIGURE 2.7. *Monthly average temperatures for each climate division of North Carolina*

mid-30s at the Outer Banks. At the opposite extreme, the average July maximum temperature (Fig. 2.8b) shows the familiar pattern, with values from 70 to 90 from west to east.

DAY-TO-DAY TEMPERATURE FLUCTUATIONS

The long-term averages say little about day-to-day conditions. A major characteristic of North Carolina's climate is the difference in day-to-day temperature fluctuations between summer and winter (see Fig. 2.9). In summer, differences are almost always small—rarely more than about 3°F. But in winter, they can be very large, and temperatures can move up or down by 20°F or more in a few hours. Usually the whole weather type, called the air mass, is involved, not just temperature. Summer may be characterized, at least outside the mountains, as a continuous daily stream of hot and humid weather, the only concern being whether it will rain or not. Winter, however, may be a seemingly constant stream of changes. One day may be cold, dry, and cloudless, to be followed by a day or so with warm, damp, cloudy conditions that, in turn, yield to another spell of cold, clear air.

TEMPERATURE EXTREMES

No discussion of temperatures is complete without consideration of temperature extremes (see Table 2.2). The record for the all-time high in North Carolina is held by Fayetteville, with 110°F on August 21, 1983. Several other stations observed their

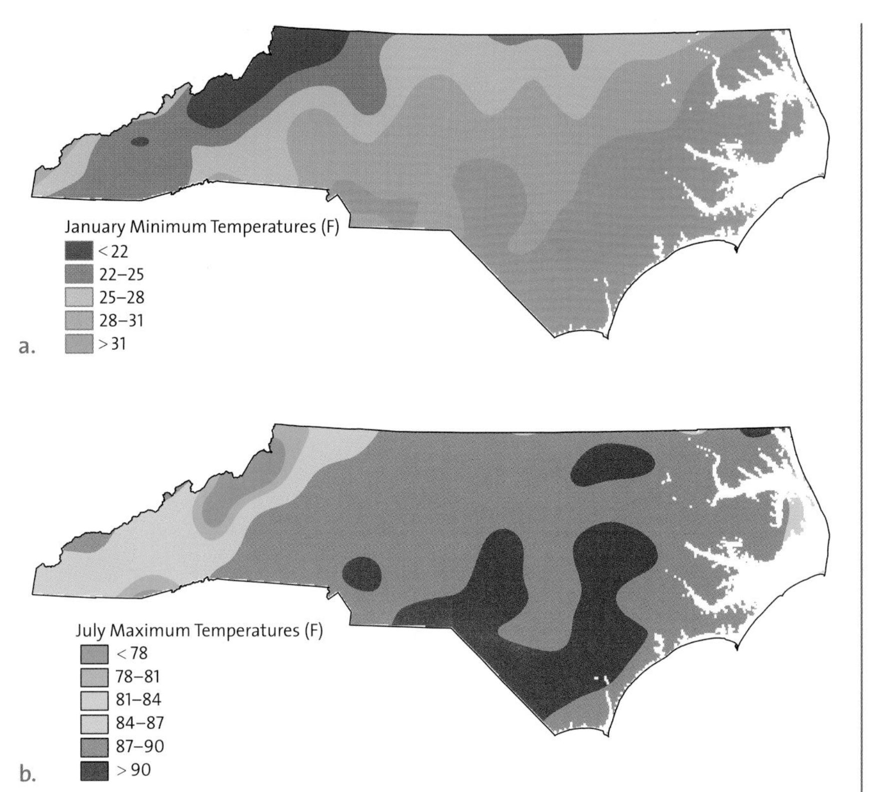

FIGURE 2.8. *Average (a) January minimum and (b) July maximum temperatures across North Carolina*

record high on or near this date. There was a similar widespread hot spell in the middle of August 1988. Both periods led to major heat waves. Records at other stations were established on a wide variety of dates, and a wide variety of extreme values were involved. In general, the all-time highs at individual stations range from the mid- to high 90s on the shore and in the mountains to around or slightly above 105°F on the Coastal Plain and in the Piedmont.

I, along with many other people, have seen electronic time/temperature displays indicating that the temperature has exceeded the state record, often by several degrees. However, all official weather observations have to be made using standard methods, instruments, and exposures (see Appendix B). For example, instruments must be shaded from the direct glare of the sun, and measurements must be made over a grassy surface. Grass is almost always cooler than tarmac, simply because it can lose energy, and therefore cool down, by evaporation. Most of the displays of high temperatures seem to me to be associated with parking lots.

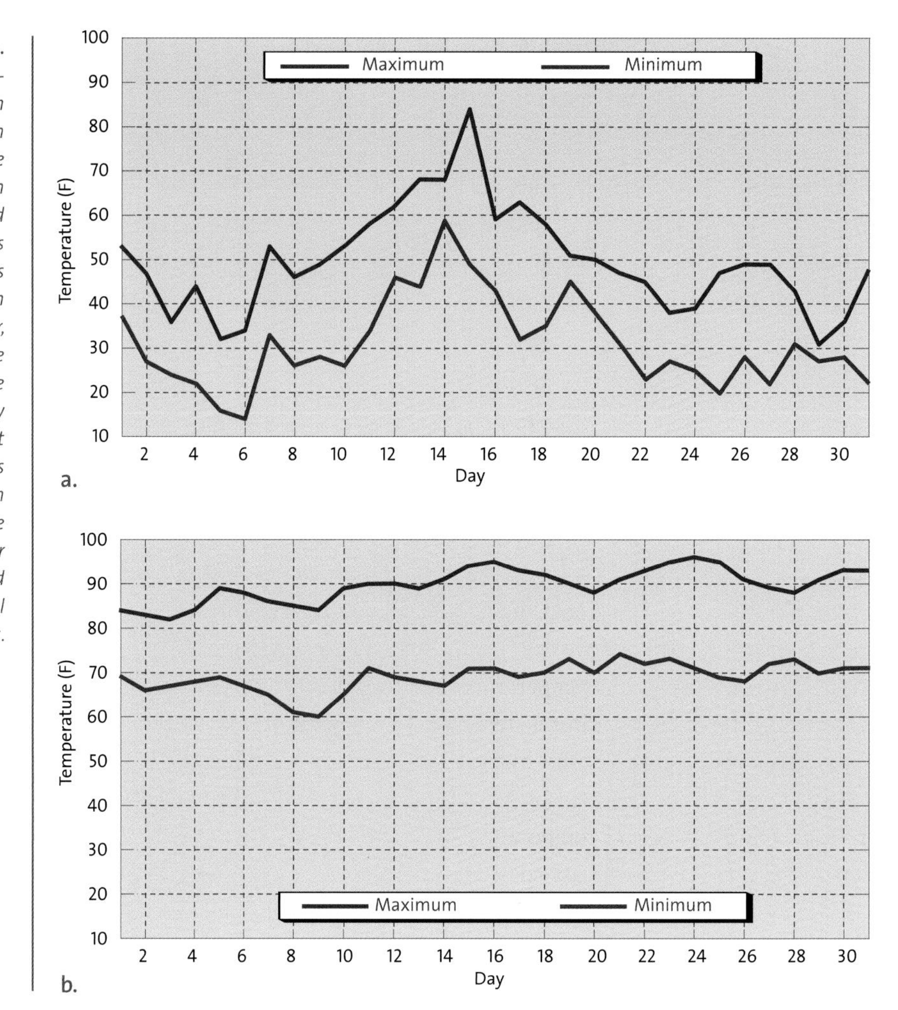

FIGURE 2.9. *Typical day-to-day maximum and minimum temperature fluctuations in (a) January and (b) July. There is much less variability in summer, although the difference between day and night temperatures is similar for both months. These data are for Piedmont Triad International Airport, 1995.*

The extremes may be the most interesting temperatures, but the frequency of hot days with temperatures above 90°F is more important for our discomfort. The statewide pattern of this frequency appears somewhat surprising at first glance (see Fig. 2.10). The fact that it is concentrated in the summer is probably not surprising, and neither is the fact that high-elevation Asheville has fewer hot days than Raleigh, which in turn has fewer hot days than coastal Wilmington. However, Cape Hatteras has very few days over 90° even in the height of summer. The low values at Hatteras are a direct result of its shoreline position. Hatteras is a rather windy place and thus stays relatively cool. In addition, wind that blows from the sea, because of the continentality effect, is cooler than wind from the land. The

TABLE 2.2. High Temperature Records for Selected Stations in North Carolina

STATION	TEMPERATURE (°F)	DATE
Asheville	100	8-21-83
Cape Hatteras	96	7-10-92
Charlotte	103	5 times, incl. 8-21-83
Elizabeth City	107	7-18-42
Grandfather Mtn.	91	8-27-68
Greensboro	103	8-18-88
Laurinburg	107	8-18-88
Murphy	99	8-23-83
Raleigh-Durham	105	7-23-52, 8-18-88
Wilmington	104	6-27-52

Wilmington station, which is at the airport, is a few miles inland, enough to cut down wind speed and remove the cooling effect of the oceanic air.

When we look at the low temperatures to establish records, we have a somewhat different problem. Temperatures decrease rapidly with altitude, so that in most cases the thermometer on the highest mountaintop sets the low temperature record. Currently that is held by Mt. Mitchell, with -34°F recorded on January 21, 1985. On that day, low temperature records were set over virtually the entire state (Box 2.1). The southeast coast from Southport to Morehead City was the exception. Record cold was recorded there—around 0°F—on December 25, 1989. This coincided with the only recent major snowstorm in the area. The presence of the snow surface undoubtedly influenced the energy balance at this time and fostered the unusually low temperatures.

Although Mt. Mitchell holds the low temperature record, it is difficult to construct a long-term climatology for the mountain. The weather station there has been moved many times over the years. Originally at the summit (see Fig. 2.11), it has been relocated several times, both to ensure that the weather did not destroy the instruments and to allow observers to read those instruments whatever the weather. The relocations thus increased the amount of shelter and reduced the altitude (by more than 300 feet). Both factors tend to increase the temperature recorded, so that long-term averages or trends are difficult to compute. As a result, we tend to use the Grandfather Mountain station for such purposes. This station was started in 1955 and since then has had only one minor relocation, in 1986, which decreased the altitude by about 6 feet. Consequently this station has a much

FIGURE 2.10. *Long-term average number of days each month with mean temperatures above 90°F for selected stations*

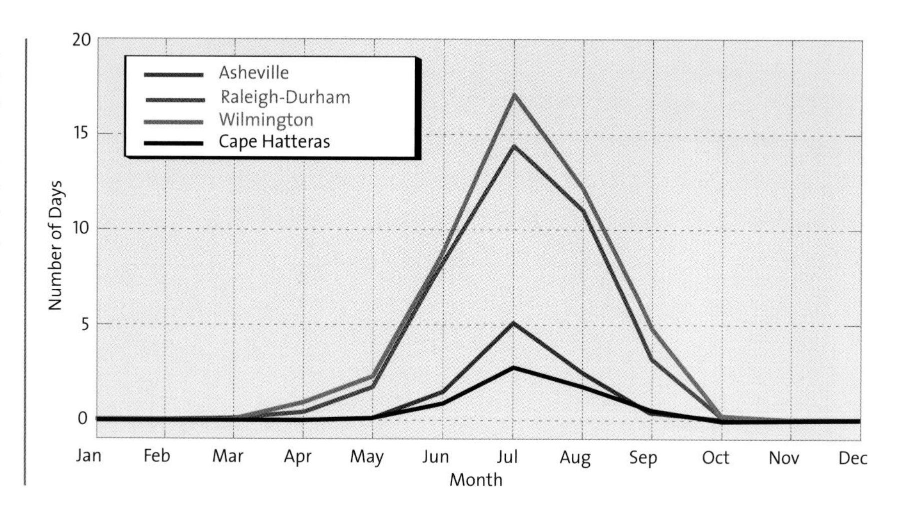

more consistent record, and we tend to use it when considering North Carolina's mountaintop climate.

Applications of Temperature Information

Knowledge of temperature is of practical use in a great variety of ways. We consider three examples here. These represent major uses we are all likely to come across fairly frequently, but they were also chosen to show the range of ways in which weather information might be of practical use in everyday life.

HEATING, COOLING, AND GROWING DEGREE DAYS

One impact of temperature that we all can appreciate is that the colder it gets, the more heating we need for our homes. Experience has found that people in the United States are most comfortable when their houses are heated to about 65°F. This might seem a little low, but "self-heating" of the house by the various motors, lights, and people within raises the temperature an extra 5° or so, giving the final comfort level. (The actual temperature we choose as comfortable depends on our lifestyle and culture—and probably on the cost of energy. Most Europeans, who pay more for energy, set their thermostats about 5°F lower than we do).

A measure of the heating needed is how much the temperature outside the house (usually called the ambient temperature) is below 65°F. This difference is the number of *heating degree days* (HDD) for the day. Days with temperatures of 65°F and above have zero heating degree days. Summation of the number for each

BOX 2.1. RECORD COLD: JANUARY 21, 1985

January 21, 1985, was inauguration day for the fortieth president of the United States. Since the maximum temperature in Washington, D.C., that day was 19°F—lower than the minimum temperature for any other inauguration—the ceremony marking the beginning of President Ronald Reagan's second term was moved indoors.

In North Carolina the night of January 20–21 set many all-time low temperature records. The whole state was involved, from north to south, from mountaintops to coastal beaches, from -32°F on Grandfather Mountain to 6°F at Cape Hatteras. In fact, much of the eastern seaboard was affected. New York City had its coldest January night since records began; Jacksonville and Daytona Beach in Florida set all-time low temperature records. Deep frost destroyed most of Florida's citrus crops. Meanwhile, places in the Alaskan interior were 20° above normal.

TABLE 2.B1. Selected Stations Where All-Time Low Temperature Records Were Established January 21, 1985

STATION	TEMPERATURE
Asheville	-16
Cape Hatteras	6
Charlotte	-5
Elizabeth City	-2
Grandfather Mtn.	-32
Greensboro	-8
Laurinburg	-3
Murphy	-16
Raleigh-Durham	-9

This extreme event was the culmination of a period during which our airflow was controlled by a ridge of high pressure along the eastern flanks of the Rocky Mountains. This ridge forced the air bringing our weather, which normally comes across the continent from the west, to take a detour northward. Warm air from the tropical Pacific was pushed north into Alaska to find a route around this blocking ridge. The warm air cooled as it continued around the north of the ridge, blowing over the cold, snow-covered land of Arctic Canada. Moving southeast, it dropped some snow as it went, eventually arriving over the east coast as a cold, dry airstream. These cold air outbreaks are common in our winter. But rarely is the flow as vigorous as that on January 21; rarely does the air start as cold, move as fast, or cross so much snow. Our air was almost unmodified from that found in the Arctic.

FIGURE 2.11. *The summit of Mt. Mitchell in 1954, showing the thermometer screen and the rain gauge just below the mountain peak (postcard published by S. H. Kress & Co.)*

day gives the total for a week, a month, or a season. In practice, we use the daily average temperature for the local weather station as the ambient temperature.

The name "heating degree days" makes no logical sense that I can determine and is probably misleading; its origin is lost in the mists of time. A better term would be simply "heating degrees," but it has never been used.

The annual total of HDD (see Fig. 2.12) around the state includes observations only for the official "heating season" between October and April. Although some places, especially in the mountains, may have some small heating requirements outside that season, the numbers are low and can safely be ignored in the annual totals. The overall pattern tends to follow that of average temperature, but with the largest number in the northwest and the smallest values in the southeast. The advantage of this map, however, is that it translates directly into heating demand, and thus roughly into the cost of heating a building in a particular area.

Maps like the one in Figure 2.12, especially when used with similar ones showing the maximum energy demand likely during the coldest midwinter periods, are useful in helping to determine the correct, most efficient size of furnace to install in a building. Indeed, many local building codes include sizing guidelines or rules that ultimately depend on the HDD information.

The summertime opposite of HDD is *cooling degree days* (CDD). This is the amount the daily average temperature is above 65°F, with days at or below 65°F having zero CDD. The "cooling season" is from May through September, and again the map of annual total CDD (Fig. 2.13) shows a distribution similar to that of summer temperatures. It can be used in the same way as the HDD map.

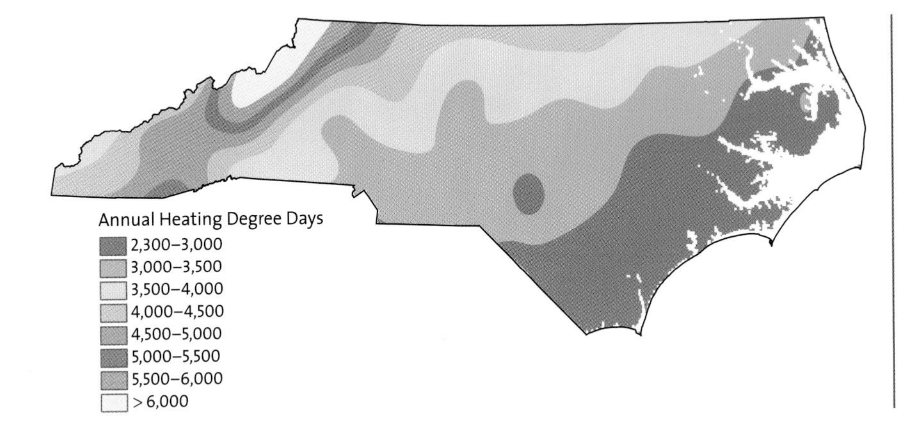

FIGURE 2.12. *Long-term averages of the total annual heating degree days across North Carolina*

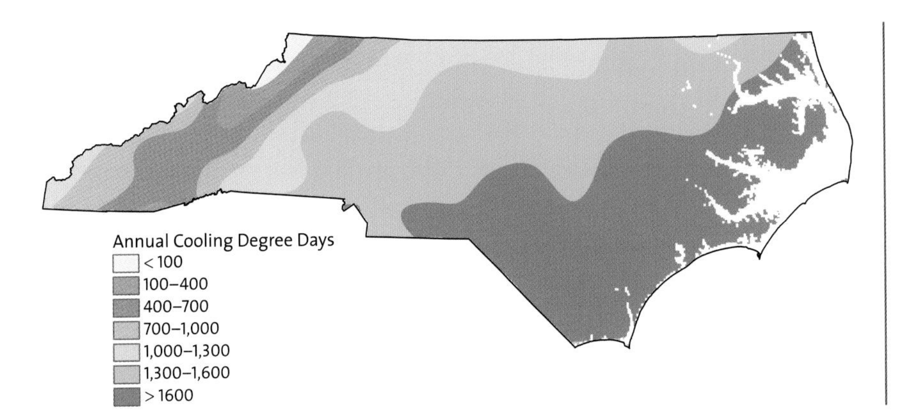

FIGURE 2.13. *Long-term averages of the total annual cooling degree days across North Carolina*

This map suggests that if you want to minimize your summer energy costs, the place to live is in the western mountains, especially at the higher elevations in the northwest. Most summer cooling seems to be needed around Morehead City and Wilmington. Figure 2.12 suggests that winter heating expenses are highest in the northwest and least around the port cities. However, since actual costs per degree day are not the same for heating and cooling, we cannot simply add the two to find the climatologically cheapest place to live.

We can, however, track our own energy usage and expense in relation to the weather. Many energy companies provide HDD or CDD information on their monthly bills. You can use this to determine how much of your monthly energy usage you can attribute directly to the weather and how much was due to your "lifestyle." If your energy company does not provide HDD information, this kind of analysis is more complicated, since you would have to track the daily temperatures of the local weather station and determine the HDD totals yourself. Even more precision could be achieved by taking the temperature measurements at your house rather than relying on the local station or the area averages that the energy company uses. Again, you would convert your measurements to HDD and relate the result to your energy bill.

Having considered our own demands for energy, we can look at those demands from the perspective of the power company. A few moments' reflection about the electrical energy usage in your home soon indicates that there are two main components. The first, more or less fixed component includes daily needs for lighting, cooking, heating water, and the like. A second, weather-sensitive component primarily reflects the temperature; the furnace works harder and longer on a cold day, or the air conditioner does the same on a hot day.

Use of energy for the first, non-weather-sensitive component may vary slightly from day to day and house to house, but taken together, all the customers of a power company tend to use virtually the same amount every day. Similarly, the mass of customers tends to respond in the same way to temperature changes. Comparison for many individual days allows the development of a clear relationship between temperature and energy demand. The value of this relationship for the power company is that it can use the weather forecast for one or a few days ahead to plan electricity generation needs, promoting efficient operation of the company's facilities.

The concept of degree days is also used in agriculture, where *growing degree days* (GDD) are commonly used. These are defined in a way similar to HDD, except that the threshold temperature depends on the temperature at which significant plant development starts, usually around 40°F rather than 65°F. There may also be an upper limit, often near 90°F, beyond which no extra GDD accumulate, since the plant growth processes shut down to prevent wilting. The most familiar application of GDD for most of us is the map that often appears on the back of seed packets indicating which regions are suitable for various varieties. Most of these are based on the GDD concept.

USING TEMPERATURE RECORDS TO FORECAST AN EVENT: THE CHANCE OF FROST

Temperature conditions vary from year to year, and it is frequently important to have an estimate of when a particular event will occur. The date of the first frost in the fall is a case in point (see Table 2.3). Forecasts can only be made a few days in advance, which is no help if you need information in the spring before making a

TABLE 2.3. Probability of the Last Frost in Spring and the First Frost in Fall on or after the Indicated Date, for Selected North Carolina Stations*

	Last in Spring			First in Fall			Growing Season Length (days)		
	PROBABILITY LEVEL (%)			PROBABILITY LEVEL (%)			PROBABILITY LEVEL (%)		
STATION	90	50	10	90	50	10	90	50	10
Andrews	4-19	5-4	5-19	09-27	10-10	10-22	180	158	136
Concord	3-20	4-2	4-16	10-17	11-2	11-18	236	213	190
Edenton	3-10	3-24	4-7	10-28	11-10	11-24	251	231	211
Hamlet	3-26	4-9	4-22	10-14	10-24	11-4	215	197	180
Kinston	3-12	3-30	4-16	10-13	10-29	11-15	240	213	186
Wilson	3-20	4-2	4-15	10-16	10-29	11-11	228	209	189
Transou	4-28	5-16	6-3	09-16	09-30	10-13	155	136	117

*Frost is defined as an overnight air temperature at or below 32°F.

decision as to what to plant and how long it will take to mature. This is a practical example of the general forecasting situation where deterministic forecasts would be desirable but are not available. Instead we have to rely on probabilistic forecasts. These are developed, using statistical techniques, from records of previous occurrences. The techniques are discussed in Chapter 7.

The numbers in Table 2.3 indicate that most of North Carolina can expect some frost after the middle of March, judging by the dates at which there is still a 90 percent probability for frost occurring in spring. Only on the Coastal Plain (Kinston and Edenton) is the 90 percent probability earlier than this. Around the middle of April, however, the chance of frost is relatively small for most of us. In the western mountains, however, there is a much greater probability, with the valley location of Andrews recording 10 percent of years with frost after May 19. The date is June 3 for Transou, high in the northwestern mountains. The dates of the first fall frost show a similar pattern, with Edenton generally being the last and Transou the first of these stations to see a frost.

Frost is defined in Table 2.3 as an overnight air temperature at or below 32°F as measured by a thermometer in a standard instrument shelter. This is some distance above the surface of the earth. The temperature difference between the shelter and the ground is not always the same. Thus there is no guarantee that an overnight temperature below freezing will bring a frost in the sense of there being a visible white coating on the ground or ice on your vehicle's windshield. Nevertheless, the likelihood of frost damage to most things that are susceptible,

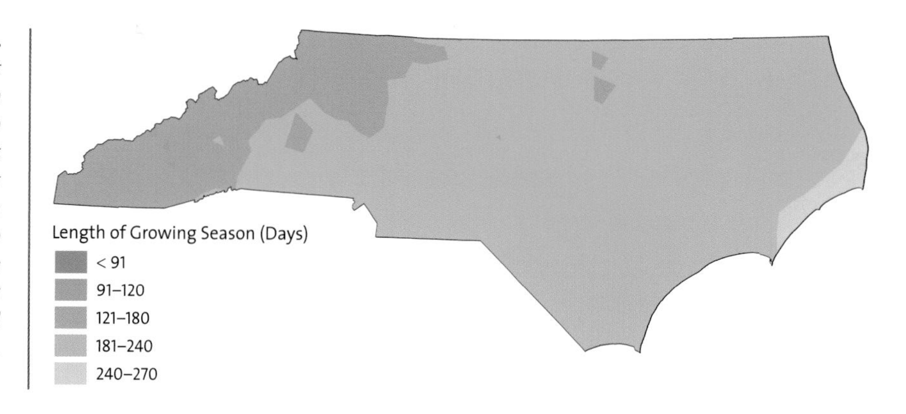

FIGURE 2.14. *Median (50 percent probability) length of the season between the last spring and the first fall frost in North Carolina. This map has much the same pattern as that showing annual temperature.*

whether agricultural crops or construction equipment, is probably best represented by this air temperature. Most vegetation can withstand temperatures somewhat below freezing because sap contains natural antifreeze. Apple and peach trees, for example, can generally withstand air temperatures down to at least 28°F before significant damage occurs. Consequently, it is often useful to look at probabilities for several different temperatures.

The length of the growing season, defined as the average time between the last frost of spring and the first frost of fall, is shown in Table 2.3 for various probabilities at various stations. For most of these there is a difference of around forty days between the shortest and the longest of the seasons. The average (50 percent probability) total length varies from around 160 days in the west to 230 days in the east (see Fig. 2.14). The interannual variation is thus large relative to the average. This has economic significance, not only in agriculture, but also in other economic sectors such as recreation and leisure.

WEATHER AND HUMAN COMFORT

We North Carolinians are proud of our ability to survive in, or even revel in, our hot, humid summers. There are times, of course, when they seem excessive, and we suffer through heat waves that can present a real danger to our health. At the other extreme, we can become uncomfortable during cold, windy winter weather, and occasionally, particularly on the high peaks, discomfort can turn into a real danger to our health and safety.

For our health and comfort we need to maintain a constant internal body temperature. If it is too high or too low, health problems, potentially culminating in death, arise. Internally the food we eat is converted by the body's metabolism into the heat required to maintain this core temperature, as well as providing the energy we need for living. At the same time, the body loses heat through the skin and lungs to the surrounding atmosphere. As the atmospheric conditions change, the metabolic rate must change to maintain our core temperature.

Physically the energy interactions between us and our atmosphere are the same as those controlling all temperatures. However, our bodies take an active role in maintaining the desirable temperature. The weather factors involved are radiation, humidity, temperature, and wind. Standing in the sun clearly adds heat to your body; this is usually a good thing in winter but is not so helpful in summer. You also gain or lose heat by terrestrial radiation exchanges with the surrounding air and objects. You will also lose energy by a convectionlike process as you expel warm air from your lungs. Conduction will provide cooling if the air temperature is cooler than that of the skin; a warmer air temperature will cause heating. As for evaporation, breathing and panting create an internal cooling, while sweating creates an external one. The evaporation rate, and hence the cooling rate, will increase as the humidity decreases or as the wind speed increases.

Most of the time our bodies can unconsciously cope with environmental conditions and we are comfortable—or at least reasonably so. However, as temperature and humidity climb, or as temperature falls and wind increases, we become increasingly uncomfortable, until the conditions become dangerous.

On the hot side, we are concerned with *heat stress*. Once the air temperature is higher than skin temperature, we cannot lose heat by conduction. If the relative humidity is also high, we have difficulty removing energy by evaporation. Our respiration will still cool us somewhat, but without precautions there is the danger that our core body temperature will rise, possibly leading to permanent health damage. To provide a measure of these possibly dangerous conditions, the National Weather Service (NWS) has developed the *heat index* (see Table 2.4). This is a mathematical combination of temperature and humidity observations designed to indicate roughly how hot the air feels to the human body. It does not, of course, take into account other factors that influence our comfort, nor can it account for our individual metabolic responses or our activities and lifestyles. But overall it provides a guide to possible health consequences (see Table 2.5).

The most extreme events are commonly called *heat waves*, although there is no official definition of that term. Using the NWS and Red Cross experience as a guide, we can assume that a heat wave is a period of at least forty-eight hours during which the daytime heat index exceeds 105 and the overnight low index does not fall below 80. Having an elevated overnight index is vital, since it means there is no respite, either directly to us or, perhaps more importantly, to our homes.

TABLE 2.4. Calculation of the Heat Index from Measurements of Air Temperature and Relative Humidity

RELATIVE HUMIDITY (%)	AIR TEMPERATURE (°F) 70	75	80	85	90	95	100	105	110	115	120
30	67	73	78	84	90	96	104	113	123	135	148
35	67	73	79	85	91	98	107	118	130	143	
40	68	74	79	86	93	101	110	123	137	151	
45	68	74	80	87	95	104	115	129	143		
50	69	75	81	88	96	107	120	135	150		
55	69	75	81	89	98	110	126	142			
60	70	76	82	90	100	114	132	149			
65	70	76	83	91	102	119	138				
70	70	77	85	93	106	124	144				
75	70	77	86	95	109	130					
80	71	78	86	97	113	136					
85	71	78	87	99	117						
90	71	79	88	102	122						
95	71	79	89	105							
100	72	80	91	108							

TABLE 2.5. Effects of Heat on the Human Body

HEAT INDEX	EFFECTS
130 or above	Heat stroke highly likely with continued exposure.
105 to 130	Heat stroke likely with prolonged exposure.
90 to 105	Heat stroke possible with prolonged exposure.

Particularly if they are not air conditioned, they never get a chance to cool down; thus the terrestrial radiation stress on our bodies increases. According to medical records, heat-related health problems do increase rapidly during and after the second day of the heat wave.

Heat waves are, fortunately, rather rare for us (see Fig. 2.15). Cherry Point Marine Corps Air Station, on the Coastal Plain inland away from the cooling ef-

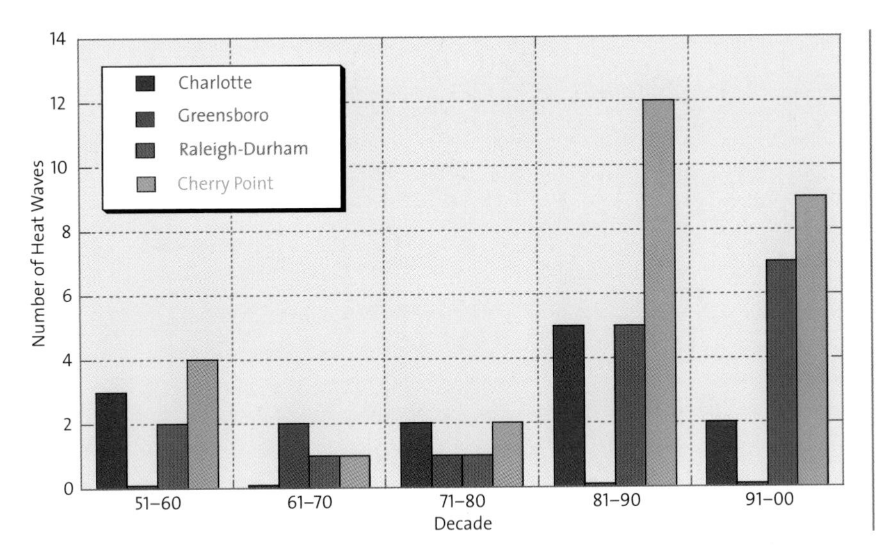

FIGURE 2.15. *Number of heat waves per decade for stations in North Carolina with fifty or more years of data. Asheville also qualifies, but no heat waves have been recorded there.*

fects of sea breezes, had several in the 1980s. Most of the Piedmont, however, gets two or three a decade, and there are few in the mountains. Heat waves seem to be localized and influence only the area around one station. A 1952 event covered most of the eastern half of the state, but that in 1999 was more typical, since it was a local Piedmont feature (Box 2.2). In most cases, however, the area outside the heat wave is unlikely to be enjoying balmy weather but, instead, hot, sticky conditions just below the heat wave threshold.

So in hot, humid conditions we should minimize heat production and maximize heat removal. To minimize internal heat production, we need to minimize activity and energy (food) intake. To minimize heat absorption, it helps to get into the shade. To maximize heat removal, we can use basic meteorological principles: drink plenty of water so we have moisture available for evaporation, seek a breezy position to increase evaporation, and wear light clothing to allow sensible heat flow from the skin.

At the other end of the weather and human comfort scale is the *wind chill*, the combination of low temperatures and high wind speeds that can also lead to major health hazards. Now the concern is that the environment fosters very rapid sensible heat transfer and evaporation. Then the warm core cannot transfer heat to the skin or any exposed extremity—fingers and toes commonly—fast enough to prevent freezing, which causes frostbite or worse. There is also the *wind chill index* (see Table 2.6), against which the NWS can issue warnings.

Most of North Carolina has relatively high temperatures and low wind speeds

BOX 2.2. HEAT WAVES IN 1952 AND 1999

Heat waves in North Carolina are of two distinct types: those that are exceptionally hot but rather dry, and those where temperature and humidity combine to create the event. That of 1952 was of the first type; that of 1999, of the second.

June and July 1952 were exceptionally hot and dry for the entire southeastern United States. The high pressure region usually centered over Bermuda became established over Georgia, preventing any local cloud formation and steering rain-bringing frontal systems away to our north. Toward the end of July a series of such systems passed on this northerly track. To their south the Bermuda high expanded more than usual and extended across Florida and into the Gulf of Mexico. A weak airflow into North Carolina from the southwest was established. This may have had its origins over the deserts of Texas and New Mexico, but it certainly was very dry, unlike our normal summer air. In addition, as it approached us, it blew over a landscape in the throes of a drought. So there was no water on the land to evaporate into the air and cool and moisten it, and the air got hotter. The result was a heat wave affecting most of the state outside the mountains. Daytime temperatures above 100°F were recorded on several consecutive days. With low humidity, temperatures did fall at night into the upper 70s. So in this case it was the heat, and not the humidity, that caused the several deaths and many illnesses reported in the newspapers.

For the heat wave of July 1999, however, it seems that it was the humidity, not the heat, that caused the problems. In many ways the weather map was similar to that for 1952, but the hot air reached us more directly from the Gulf of Mexico. The southeastern states were also much wetter, so it was a hot, humid wind that arrived. The consequence was, again, hot days. But the humid air ensured that temperatures did not fall much during the night, and we humans continued to sweat and suffer. Probably because of what appeared to be small variations in humidity, this heat wave did not cover the whole of the eastern part of the state. It seemed to occur in spots, picking out the Triangle and much of the central coast but just missing the Triad and the southern Piedmont. Even so, it led to at least two deaths and caused health problems for many of the very young and very old.

in winter so that none of the large population centers are likely to experience major problems from the cold equivalent of a heat wave. This is not to say that we cannot experience cold conditions, and we need to dress and act appropriately to avoid unnecessary prolonged exposure. But only on exposed mountain peaks are conditions likely to become severe. The only meteorological advice following a forecast of such conditions is to dress very warmly to cut down loss of sensible heat from the skin, keep dry to minimize evaporation, and seek shelter out of the wind. In most cases getting to a lower elevation should be the prime objective.

Mt. Mitchell and similar high peaks can be exceptionally cold in winter, and

TABLE 2.6. Wind Chill Calculations Based on the Understanding of Human Thermal Comfort Devised in 2001

WIND SPEED (MPH)

TEMPERATURE (°F)

0	5	10	15	20	25	30	35	40	45	50	55	60
40	36	34	32	30	29	28	28	27	26	26	25	25
35	31	27	25	24	23	22	21	20	19	19	18	17
30	25	21	19	17	16	15	14	13	12	12	11	10
25	19	15	13	11	9	8	7	6	5	4	4	3
20	13	9	6	4	3	1	0	-1	-2	-3	-3	-4
15	7	3	0	-2	-4	-5	-7	-8	-9	-10	-11	-11
10	1	-4	-7	-9	-11	-12	-14	-15	-16	-17	**-18**	**-19**
5	-5	-10	-13	-15	-17	**-19**	**-21**	**-22**	**-23**	**-24**	**-25**	**-26**
0	-11	-16	**-19**	**-22**	**-24**	**-26**	**-27**	**-29**	**-30**	**-31**	**-32**	**-33**
-5	-16	**-22**	**-26**	**-29**	**-31**	**-33**	**-34**	**-36**	**-37**	**-38**	**-39**	**-40**
-10	**-22**	**-28**	**-32**	**-35**	**-37**	**-39**	**-41**	**-43**	**-44**	**-45**	**-46**	**-48**
-15	**-28**	**-35**	**-39**	**-42**	**-44**	**-46**	**-48**	**-50**	**-51**	**-52**	**-54**	**-55**
-20	**-34**	**-41**	**-45**	**-48**	**-51**	**-53**	**-55**	**-57**	**-58**	**-60**	**-61**	**-62**
-25	**-40**	**-47**	**-51**	**-55**	**-58**	**-60**	**-62**	**-64**	**-65**	**-67**	**-68**	**-69**
-30	**-46**	**-53**	**-58**	**-61**	**-64**	**-67**	**-69**	**-71**	**-72**	**-74**	**-75**	**-76**
-35	**-52**	**-59**	**-64**	**-68**	**-71**	**-73**	**-76**	**-78**	**-79**	**-81**	**-82**	**-84**
-40	**-57**	**-66**	**-71**	**-74**	**-78**	**-80**	**-82**	**-84**	**-86**	**-88**	**-89**	**-91**
-45	**-63**	**-72**	**-77**	**-81**	**-84**	**-87**	**-89**	**-91**	**-93**	**-95**	**-97**	**-98**

Frostbite occurs in 15 minutes or less.

Wind chill (°F) = $35.74 + 0.6215T - 35.75(V^{0.16}) + 0.4275T(V^{0.16})$,
where T = air temperature (°F) and V = wind speed (mph).

parts of the Coastal Plain can be exceptionally hot in summer. We may experience discomfort then, but for most of the time our state has benign temperatures. There are differences from place to place on any day and distinct seasonal contrasts. These can certainly be explained in meteorological terms, but they also provide a stimulating and invigorating environment for us.

chapter 3

Water in the Air and at the Ground

Clouds and precipitation, along with temperature, are probably the most familiar elements of weather and climate. Everyone in North Carolina has probably said at some time, "It's not the heat, it's the humidity," or has worried about wilting plants in fields or gardens. All of these are aspects of water in the air and at the ground and have practical effects on our lives. They are linked by the hydrological cycle (Fig. 3.1). In this chapter we follow water around the cycle, looking at the causes and consequences of the movement through evaporation and humidity to clouds and rain and finally to river flow and water supply.

The Flow of Moisture into the Air

Evaporation tends to be, for most of us, a silent, almost unnoticed component of the hydrological cycle. We might consider it in passing if we dry clothes on an outside line, or we might express concern in the early afternoon of a sunny July day when our tomato plants begin to wilt. If we are determining the irrigation needs of an agricultural crop or monitoring evaporative losses from a town's water reservoir, a detailed knowledge of evaporation rates is vital. Indeed, one of the major characteristics of the North Carolina climate—and one with major practical consequences—is the difference between the amount of precipitation that fills our lakes and reservoirs or enters our soil and the amount of water evaporated back into the atmosphere. In some years the rain input dominates, and we have plenty of water. In others, evaporation is greater than rainfall, and drought occurs.

Evaporation is the process by which liquid water is transformed into vapor and is transferred from the ground to the air. It can occur in two ways. First, there is the direct transfer from an open water surface. This is *evaporation*. The open water surface may be as large as the Pacific Ocean or as small as a drop of water

FIGURE 3.1. *Schematic diagram of the hydrological cycle*

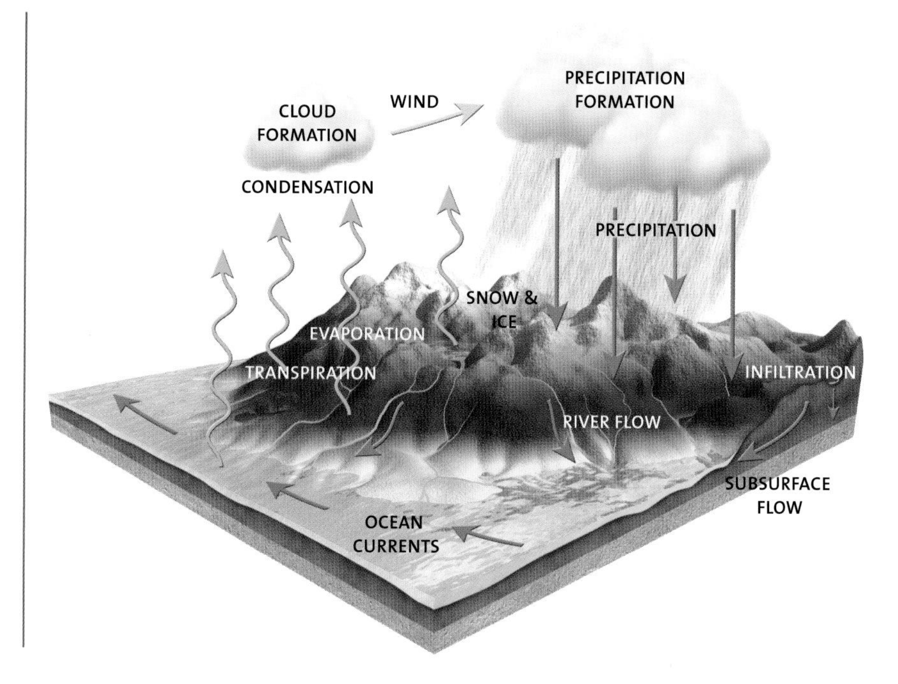

between soil particles; the process is the same. The second process is *transpiration*. Living plants extract water from the soil through their roots, use it in their various life processes, and eventually eject it as a vapor into the atmosphere through the stomata on their leaves. Most of the time we deal with both together, and the term *evapotranspiration* has been coined for the combination.

FACTORS CONTROLLING MOISTURE FLOWS

The rate of evapotranspiration—and thus the amount of water evapotranspired over a given time—depends on three atmospheric conditions: energy availability, wind speed, and atmospheric humidity. The needed energy is that required to create the latent heat flow we considered in the last chapter. As long as there is more incoming than outgoing energy, we can have evaporation. Although this means that some evaporation can occur at night, by far the most important factor is the amount of sunshine. The rate of evapotranspiration strongly depends on solar radiation. The rate also depends on the moisture gradient, the rate of change in moisture amount as we move upward from the earth's surface. The drier the air above the surface—that is, the lower the atmospheric humidity—the faster the evaporation. The role of the wind is to maintain a steep moisture gradient. In calm conditions the moisture evaporated from a water surface stays close to the surface, the near-surface air layer becomes saturated, and evaporation slows. With wind there is turbulent mixing, and the wet surface layer is carried upward and mixed with drier air. The faster the wind, the greater the mixing and the greater the evaporation rate.

As long as water is available, these three atmospheric conditions alone control the evapotranspiration rate. This is the *potential evapotranspiration*. It is a measure of (1) water lost from a farm pond or a city reservoir or (2) water used by plants in a field adequately supplied with water.

Water is not always readily available, however. Over many rural surfaces there are times, notably on a summer afternoon, when the soil and plants are incapable of moving water to the surface or the roots as fast as the atmosphere can cause evaporation. Soil surfaces begin to dry out, and plants begin to wilt. Their rate of evapotranspiration, unsurprisingly called the *actual evapotranspiration*, will be less than the potential rate. Although many plants benefit from short periods of moisture stress when the actual evapotranspiration is less than the potential evapotranspiration, prolonged stress is usually detrimental. Since in most North Carolina summers such prolonged stress is common, irrigation is needed to maintain the actual evapotranspiration close to the potential rate.

Urban areas, in contrast to rural ones, tend to evaporate water either at the potential rate or not at all. A city street is either wet or dry. During and immediately after rain, the water is being evaporated at the potential rate or drained away underground. Once a street is dry, there is no evaporation. As a result, energy that would have been used for evaporation goes into surface heating, and cities become warmer than their surroundings.

EVAPORATION IN NORTH CAROLINA

We have some direct measurements of evaporation; but they are few and scattered, and most stations have been in operation for only a few years. Nevertheless, the results from four long-term evaporation stations give some idea of the amount of water moved into the atmosphere (see Fig. 3.2). Generally the Coastal Plain, represented by Aurora, has the greatest annual total; the mountains (Coweeta) have the least; and the Piedmont (Chapel Hill) lies somewhere in between. The results for Hofmann Forest, which is on the Coastal Plain but has results similar to Chapel Hill's, indicate the uncertainty in our measurements. However, the seasonal pattern is similar for all stations, with the most water loss in July, although Coweeta in June has almost as much loss as in July. All four stations show mini-

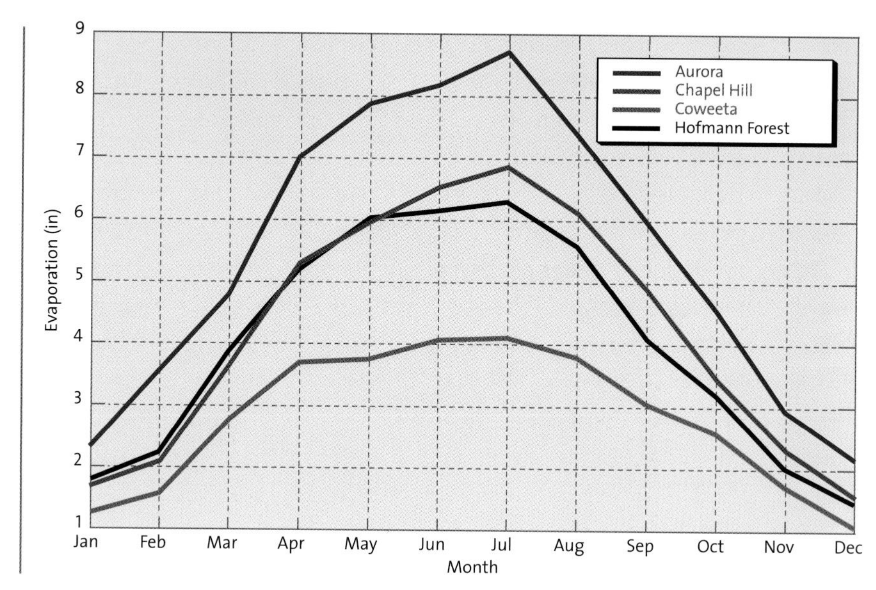

FIGURE 3.2. *Long-term average monthly pan evaporation for four North Carolina stations*

mum evaporation in December and January. For evaporation values away from these sites, we have to estimate values from measurements of temperature, radiation, humidity, and wind speed.

Humidity in Our Air

Humidity in North Carolina is almost synonymous with summer. Indeed, as a factor in our own comfort, it is a very obvious aspect of climate. In the hydrological cycle it plays a role in evapotranspiration and is vital as a precursor of clouds.

HOW DO WE EXPRESS HUMIDITY AMOUNTS?

There is no one way to express the amount of moisture in the atmosphere that will cover all our needs. However, since in practical terms we are concerned with the human perception of moisture and with cloud formation, we shall consider only two factors: dew point temperature and relative humidity. Since these depend on the concept of saturation, we have to explain this first.

Saturation represents the upper limit to the amount of moisture that the air can hold at a particular temperature. If we try to evaporate more water into saturated air, the air will start to condense that water vapor back to liquid water, ensuring that the saturation amount is not exceeded. Most of us are familiar with this concept: if we take a long shower, a fog begins to form in the room. Shower water originally simply evaporated into the air. This increased the humidity above saturation, and so condensation took over to create the liquid water droplets we see as fog or cloud. This works fastest—and we are most likely to see fog—with a hot shower in a cold bathroom. Sometimes the water vapor condenses onto the walls and ceiling rather than into the air itself, giving us dew rather than steam.

To get our useful measures of the moisture in the air, we use saturation in two ways. First we can determine the *dew point*. It is defined as the temperature to which the air must be cooled, without change in moisture content, for saturation to occur. Think of summertime when you take a cooled drink can out of the refrigerator and leave it on the kitchen counter for a few minutes. Provided the air in the kitchen is fairly calm, the air near the cold can will be cooled through contact with the can. Eventually it will be cooled below the dew point, and dew will form. We see it as condensation on the can. If we measured the temperature of the air in contact with the can at the exact instant the dew formed, we would have measured the dew point. This is easy to say but not easy to do. Appendix B explains how the National Weather Service measures humidity. The dew point, although expressed as a temperature, is actually a measure of the amount of water vapor in the air. This may seem confusing, or even perverse, but it has great utility in weather forecasting.

Our second humidity measure is the *relative humidity*. This is the ratio of the amount of moisture that the air is actually holding to that which it could hold before becoming saturated at the current air temperature. It is often taken to be a measure of how close the air is to saturation. Traditionally this is the measure that has been most commonly quoted in the public weather forecasts. In general it is the best measure for estimating the "feel" of the moisture of the air. Since we humans sweat, our skin is usually close to saturation, and the relative humidity approximates the humidity gradient away from the skin. In turn this influences the evaporation rate for the sweat and the cooling effect of the air. While relative humidity is a good humidity measure from the human perspective, the major problem meteorologically is that it depends on both temperature and humidity. Figure 3.3 illustrates this. On this rather calm, sunny summer day the actual amount of moisture in the air, indicated by the dew point, changed very little. There was a slight increase in the afternoon as the moisture evaporated by the surface added to the vapor in the near-surface layer. However, the air temperature increased rapidly from dawn until the middle of the afternoon and then decreased. The relative humidity tracked this temperature change, not the humidity condition. Generally, relative humidity will be high at dawn, low during the middle of the day, and will

FIGURE 3.3. *Diurnal pattern of dew point temperature and relative humidity at Greensboro on July 4, 1995*

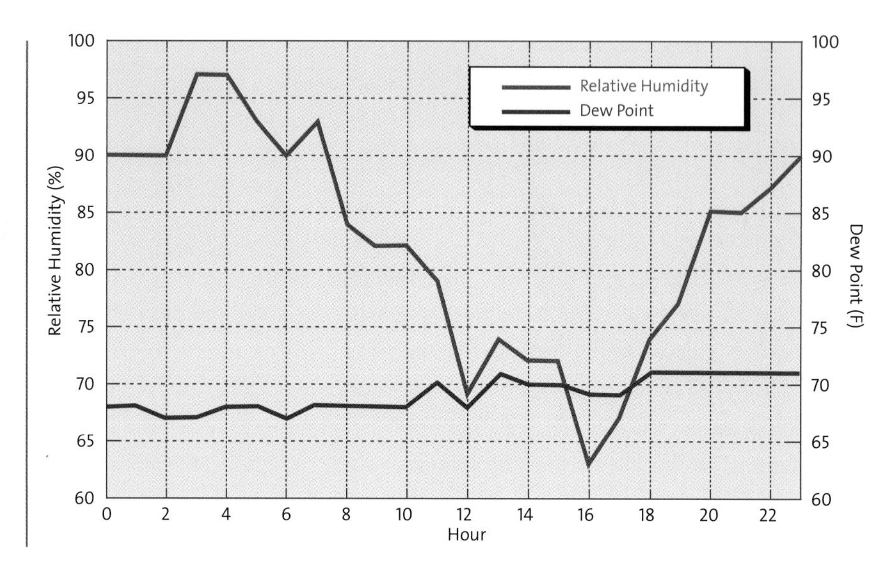

rise during the night. Also note that the air is probably not at 100 percent relative humidity when rain is falling. The cloud above will be at 100 percent, but the near-surface air need not be even close to saturation.

HUMIDITY IN NORTH CAROLINA

The moisture in the North Carolina atmosphere is a mixture of water from our own local evaporation and water that is blown into our area from other regions. For both, evapotranspiration reaches a maximum in summer, and so does the dew point (see Fig. 3.4). Cape Hatteras, close to a major water source, has the highest values. Once we leave the coast, however, there is not a great deal of difference across the state. Primarily this occurs because the atmosphere is well mixed. Locally evaporated moisture may dominate on calm days, giving a variety of wet or dry spots across the state. But for most days the winds mix the local and imported vapor to give an even distribution.

For most of the year both the local surface and the surrounding land and ocean have plenty of moisture available for evaporation, and so, relative to many other areas, North Carolina is a humid state. We notice this most in the summer, when our humidity is on a par with that of Florida (see Table 3.1). It may appear surprising that Phoenix has a higher value—more moisture in the air—than Seattle. However, the air temperature in Arizona is much higher than that in Washington State, so that the relative humidity is much lower, and the air feels much drier in the desert state.

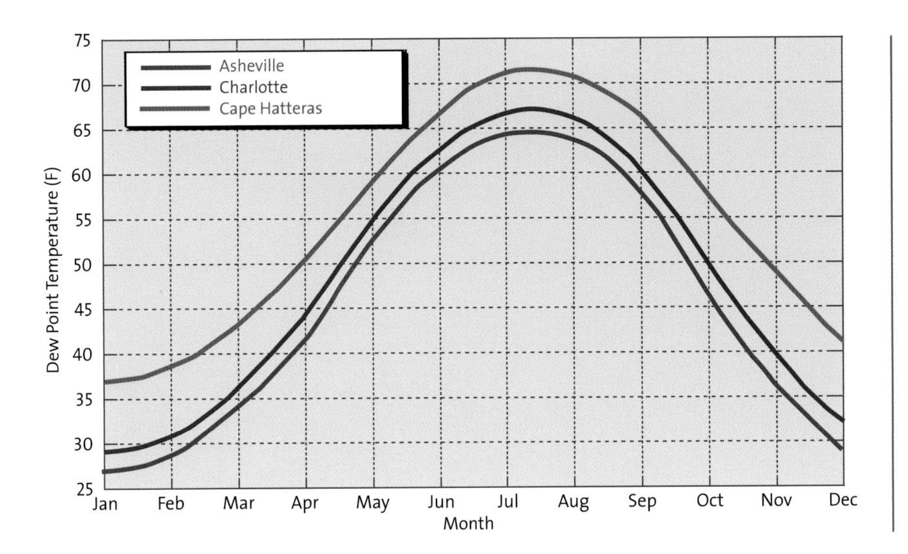

FIGURE 3.4. *Monthly average dew point temperatures for three North Carolina stations*

Clouds

Clouds form whenever the air is cooled to or below its dew point. For almost all clouds this cooling is associated with vertical motions in the atmosphere. However, fog is also a type of cloud, and it is commonly created without vertical motions. We will look at fog first and then consider the various cloud types and the processes that form them.

CLOUD AT THE GROUND: FOG

Fog, as with any cloud, consists of tiny droplets of water suspended in the atmosphere. For most of us the main characteristic of fog is the resultant reduced visibility. The various types of fog are defined by the amount of the reduction. *Fog* occurs when visibility is less than 1 kilometer (0.62 miles), while *heavy fog* has visibility less than 0.25 miles. *Mist* is less rigorously defined but refers to any conditions where visibility is obscured but where it is possible to see farther than 1 kilometer.

With our moist North Carolina climate, we are most likely to encounter *radiative fog*. This forms on clear, calm nights. The earth's surface and the air just above it then cool because of the loss of energy by terrestrial radiation. If the cooling is sufficient to drop the near-surface temperature below the dew point, fog formation starts. This often occurs soon after nightfall, and the fog gets thicker and deeper throughout the night. Soon after dawn, once the sun is up, solar radiation

TABLE 3.1. Comparison of July Average Dew Points for Selected North Carolina and Other U.S. Stations

STATION	DEW POINT
Asheville	65
Charlotte	67
Cape Hatteras	72
Miami	73
Phoenix	57
Seattle	52
Boston	61

warms and evaporates the fog. As the morning progresses, the fog slowly dissipates, rarely persisting beyond noon. A second kind of fog, *advection fog*, can occur when a warm, moist airstream blows over a cooler surface. Energy is transferred downward from the air, thus cooling it. If it is cooled below its dew point, fog will form. These fogs can form at any time and anywhere in our state, but they are most common in winter. Often in that season an airstream that has been traveling over the warm waters of the tropical Atlantic Ocean blows onshore. Contact with the cold land surface soon chills the air below its dew point, and fog forms. This fog, starting somewhat inland of the coastline, can penetrate far inland and may persist as a heavy fog for many hours. In addition, thin layers of mist, usually too shallow to cause problems, often form over many of our lakes and rivers (see Fig. 3.5).

Along the North Carolina coastline, represented by Cape Hatteras, winds often provide turbulent air mixing that discourages fog formation, so there are few heavy fogs at any time of the year (see Fig. 3.6). The conditions in Charlotte are typical of both the inland Coastal Plain and the whole Piedmont. A few radiative fogs occur throughout the year, while the addition of advective fogs gives the overall winter maximum. The frequency of fog in the mountains depends greatly on topography. The Asheville station is located in a bowl in the hills, and the position encourages cold air to drain downslope and collect in the valley during the night. This encourages the development of a radiative fog, leading to the high summer frequency. In contrast, it seems likely that most hillsides or mountaintops have very few fogs, although the problem on the mountains may be that they are in actual clouds.

We also see a third kind of fog, *evaporative fog*. This is rarely thick enough to become more than a mist. It often occurs on a calm evening over water or a very

FIGURE 3.5. *Shallow fog on a Swain County lake. This formed on a calm summer evening when the air temperature fell much more rapidly than that of the lake waters. The air was kept moist by continued evaporation from the lake, and the air eventually cooled below its dew point.*

wet land surface. The air temperature drops faster than that of the water. The water, warmed during the preceding sunny day, continues to evaporate moisture into the air. Calm conditions ensure that the air is not stirred, so that the moisture content increases while the temperature drops. When condensation occurs, we see the water drops as a wispy mist, and the surface looks as if it is smoking. The drops are, in fact, moving upward into the drier air above. Eventually water vapor entering the air from the slowly cooling water surface will be less than that carried upward, the foggy air layer will become unsaturated, and the fog will dissipate.

CLOUD TYPES WE COMMONLY OBSERVE

Like fog, clouds are created by the cooling of air below its dew point. Every cloud is unique in appearance, but we can divide them into some general classes (see

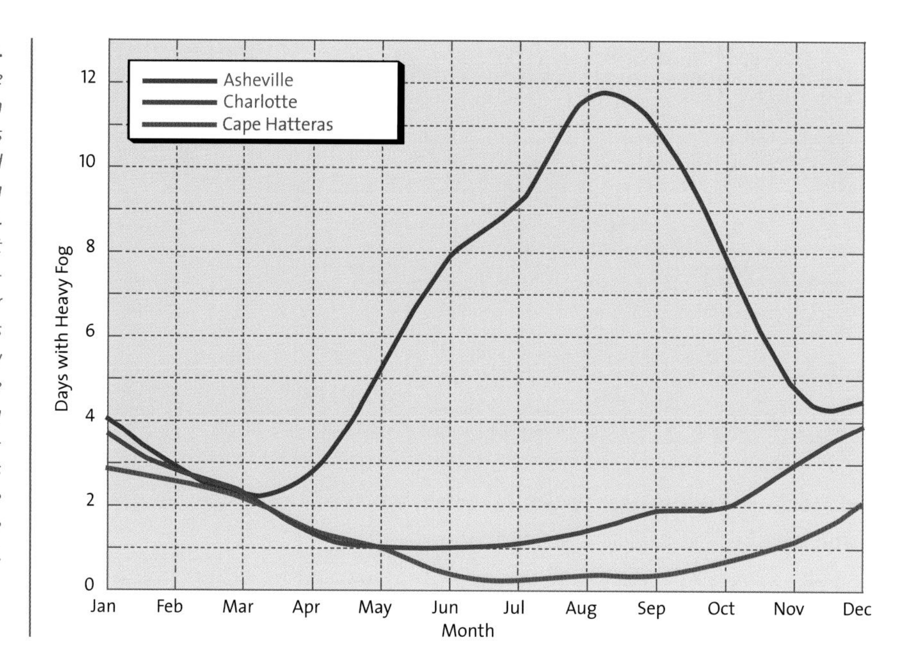

FIGURE 3.6. *Monthly average number of days on which heavy fog is reported at selected North Carolina observing stations. The values do not consider the duration of the fog. For summer conditions in Asheville, many days will have short-lived morning fog. At other seasons and stations the fog may be much more persistent.*

Fig. 3.7). Two classes, *stratus* and *cirrus*, represent clouds with mainly horizontal development. The former are low-level horizontal clouds, while the latter form at a much higher level and look much more wispy. Another class, *cumulus*, is for vertical clouds. The final category, *alto-*, refers to middle-level clouds, either high forms of stratus or cumulus clouds that start in the middle, rather than the lower, atmosphere. Some of the various classes frequently occur in combination, and Figure 3.7 shows schematically the types we are likely to see.

HOW CLOUDS ARE FORMED

Rising air cools. If it cools enough, it will reach its dew point, and a cloud will begin to form. The type of cloud created will depend on how the upward air movement was triggered. There are four main ways, and we see them all in North Carolina (see Fig. 3.8).

When air is forced to rise, the process is known as *orographic uplift*. The North Carolina mountains are a major cause of such uplift. The air slides up the mountainside, cooling as it ascends. Eventually it cools below the dew point, and a cloud forms. In many instances precipitation—*orographic precipitation*—occurs (see Fig. 1.2). Once it passes over the mountaintop, the descending air warms, the

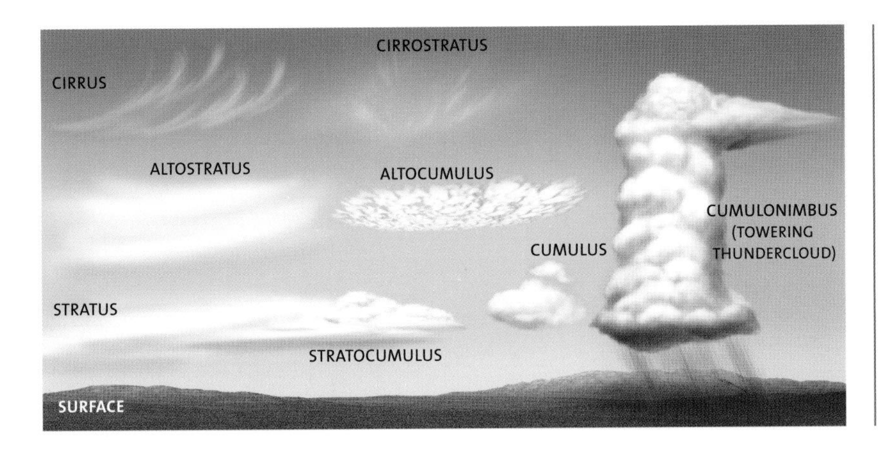

FIGURE 3.7. *Major types of clouds, grouped according to their general appearance*

cloud droplets evaporate back into the air, and the cloud dissipates. This dry, lee side is in the *rain shadow* of the mountains.

The orographic effects of our western mountains are very unusual. Most mountain ranges have airflow almost exclusively from one direction, so that the upwind side is wet day after day, while the lee side is continuously dry and may even be a desert. The Rocky Mountains of the western states are a good example. The Appalachians have no such preferred wind direction. One day the wind may come from the west, giving rain on the Tennessee-facing sides of the hills. The next day the wind may be coming from the east, and those same hillsides will be dry while the slopes facing the North Carolina Piedmont become the wet ones. Only the mountain basins are more or less guaranteed a rain shadow. The Asheville basin is the best known. There is still plenty of cloud—even if it is on the distant hillsides—but these basins are some of the driest places in the state.

Although the orographic effect is greatest in the Appalachians, any slope can give an orographic effect. Very humid air arriving over our state from the south has blown in off the Gulf Stream and is likely to be warm and very moist. Even the small rise as it crosses the Coastal Plain and enters the Piedmont may be enough to create a cloud. In this case the gentle slope spawns a stratus cloud, with its more horizontal and extended appearance, rather than the vertical cumulus clouds associated with the mountain effects.

The other cloud-forming processes are not as closely linked to specific geographic locations. Indeed, the second process, *convection*, is more closely related to the seasons. In its simplest form convection is very similar to what happens when you put a pan of water on the stove to boil. Heat from below causes the bottom water to become less dense than the overlying layers, the deep water moves upward, and

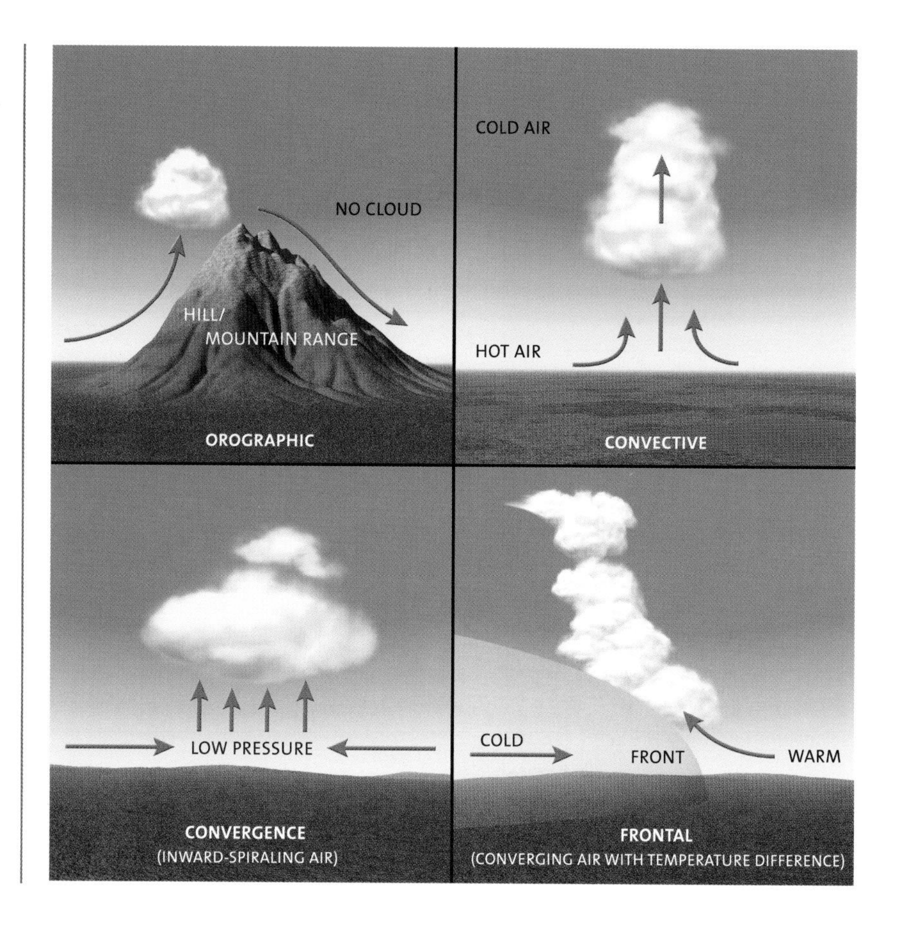

FIGURE 3.8. *Main processes creating clouds. All involve forcing humid air to rise, cool, and condense. Each process may act alone, but it is common for two, or sometimes more, to act together.*

colder water replaces it at the surface, setting up a convection current. Because air, unlike water, can be squashed, we have to take into account air pressure differences as well as temperature, but the end result is the same. We commonly see convection occur on what starts out as a cloudless summer morning. As the day progresses, more solar energy is absorbed at the ground. The ground heats and eventually becomes warm enough to start an upward convection current. Small cumulus clouds develop. During the afternoon, succeedingly bigger clouds may occur, eventually leading to a cumulonimbus cloud and a late afternoon thunderstorm. However, by evening the ground is beginning to cool, and there is no longer the energy available to keep the convection going. The clouds vanish as the droplets evaporate back into the air.

Such spontaneous uplift and cloud formation is common over our state in the summer, but convection can also be triggered when uplift is created by the orographic effect or by air convergence. In these cases, the cumulus clouds normally associated with convection are often mixed with other types of clouds, creating complex cloud patterns. These processes can go on at any season and time of day, not just on a summer afternoon.

Air convergence leads to two cloud formation processes. First, common in our area, is an inward-spiraling airstream that slowly squashes the air in the center, causing a widespread, usually rather gentle *cyclonic uplift* and an area of stratus cloud. At other times the convergence is more closely associated with two distinct airstreams of different temperatures sliding together. They meet at a weather *front*, and the warmer, less dense airstream rises over the colder, more dense one. This is *frontal uplift*. There are various types of fronts, and a range of cloud types is associated with them. Both cyclonic and frontal uplift are parts of the wave cyclones that produce much of our rainy weather. Indeed, the wave cyclones control much of our daily weather. We will examine them in detail in the next chapter.

Regardless of the process that causes uplift at a particular time and place, the cooling occurs at a fixed, known rate. To the weather forecaster, one advantage of using dew point to express moisture is that simple calculations can be used to determine how much lifting is needed to create clouds. If, for example, we know the temperature and dew point of an airstream coming onshore at Wilmington, we can determine how much it has to rise before clouds can occur. That allows us to make an estimate of how far it must travel inland—and therefore upward over the Coastal Plain—before condensation starts. So we can determine not just where, but when the cloud will form.

CLOUD AMOUNTS ABOVE OUR STATE

We do not have long-term records of the occurrence of particular cloud types or the processes that create them. Rather, we have records of the total cloud amount. This is a visual observation, with the observer estimating the amount of the sky covered by clouds, usually to the nearest eighth (okta), but sometimes to the tenth, of sky covered. On many days there may be more than one cloud-forming process acting, and more than one cloud layer. This makes precise determination of cloud amount difficult. Nevertheless, if we consider only total amount, the results are similar throughout the state (see Fig. 3.9). For much of the year, on average, about half of the sky is covered. Fall has markedly less cloud cover. It is also the season with the least precipitation and often the most sunshine.

FIGURE 3.9. *Monthly average sunrise-to-sunset cloud amounts (oktas) for selected stations*

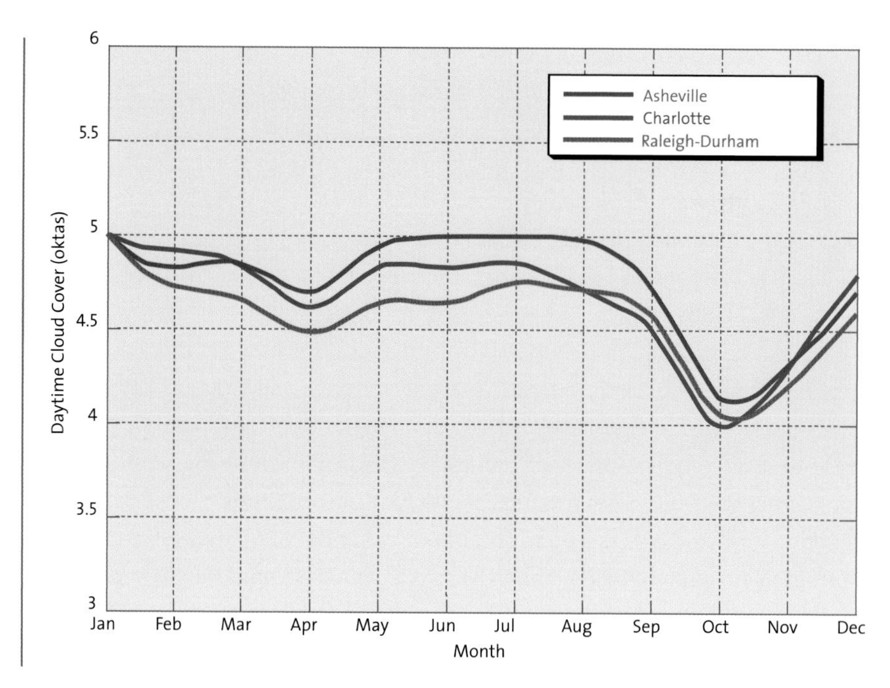

Precipitation

GETTING CLOUDS TO GIVE RAIN

Not all clouds give rain. Some form, float across the sky, and then dissipate. These clouds contain a multitude of small water droplets, created directly by condensation, that are too small to fall on their own and never grow into raindrops. Other clouds, however, contain a mix of droplet sizes or a mix of water droplets and ice crystals, and the mix allows some of the small droplets to grow into big drops. The amount of growth needed to create raindrops depends on the type of cloud. Rising, cooling air is needed for any cloud formation, so that a raindrop has to be big enough to fall through rising air on its way to the ground. The bigger the drop, the faster it can fall. So if we have a gently rising layer of stratus clouds, the drop does not have to be very big before it falls out. Stratus clouds are usually associated with light rain or drizzle. A cumulonimbus, on the other hand, is built by air rushing upward. Big drops are needed, and we associate thunderstorms with short-lived but intense rain with big drops.

Even when a drop is big enough to fall out of the cloud, it must pass through the atmosphere before we get rain. That atmosphere, by the very nature of cloud formation, is not saturated. So some evaporation must occur as the drop falls. In

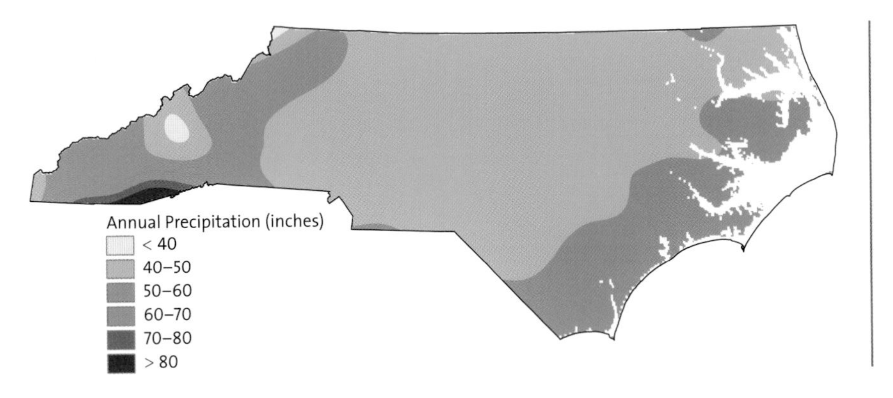

FIGURE 3.10. *Annual total precipitation across North Carolina*

some cases the drop will evaporate completely before it hits the ground, and we receive no rain. Sometimes we can see this happening when the sun shines through the falling rain and produces often spectacular shafts of light, called *virga*, below a cloud.

PRECIPITATION AMOUNTS AND DISTRIBUTION IN NORTH CAROLINA

Annual total amounts of precipitation over our state vary from around 45" in the Piedmont to more than 80" in parts of the mountains (see Fig. 3.10). In this latter region there is a great deal of spatial variability. The wettest areas are those where orographic precipitation dominates. Nearby rain shadow locations may have much lower values. The coast is much more uniform but is also wetter than the Piedmont. This is mainly because of precipitation from coastal storms, although clouds and rain associated with sea breezes have an influence.

The precipitation distribution across the state for individual months is similar to that for the annual total. Again, the west and east tend to be wet, while the Piedmont is relatively dry. In summer the whole Coastal Plain is wet, but the areas around Wilmington have the highest values. The difference is not as great in winter (see Fig. 1.7).

All areas of the state receive precipitation every month. There is relatively little difference between the seasons. The current data generally suggest that spring is the driest season and summer the wettest, especially on the coast (see Fig. 3.11). The interseason differences are minor. In the 1950s the averages indicated that fall was the wettest season, although not by much. At that time hurricanes, with their large rainfall amounts, were more common than now, and their extra rainfall contribution made the difference.

FIGURE 3.11. *Annual cycle of precipitation as indicated by the monthly total amounts for various stations*

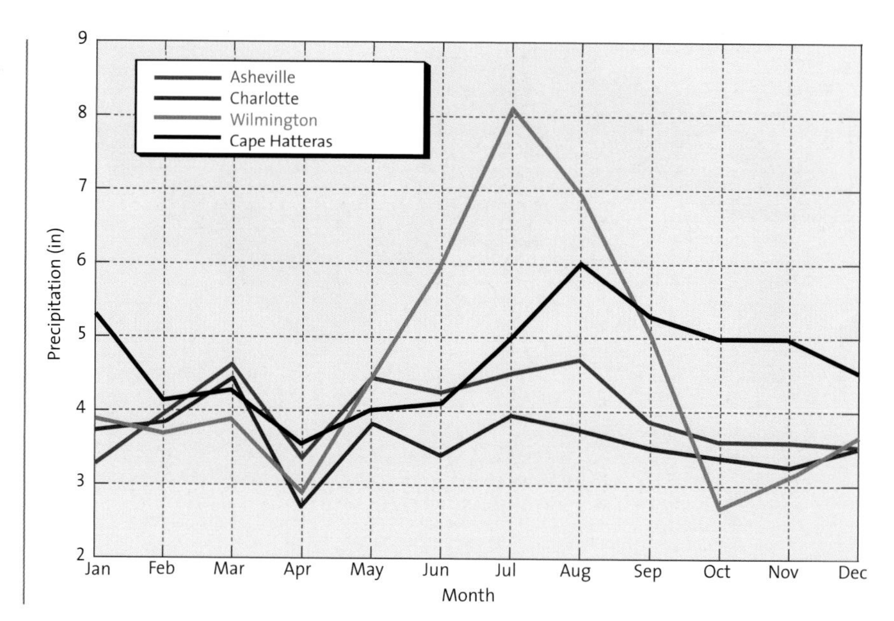

Although we usually think of precipitation in terms of the amount that falls, it is often more useful to know the number of days with rain. A *rain day* is defined as a day with 0.01" or more of precipitation. That happens to be the smallest amount of rain we can reliably measure. The annual total number of rain days varies across the state from around 115–20 on the coast to just near 110 in the Piedmont and more than 120 in the mountains. In general, therefore, it means that we shall have rain on about one day in three. Looking at monthly numbers, there is a general maximum in the summer and a minimum in the fall (see Fig. 3.12). Comparing this number with the average monthly amount of precipitation indicates that there is a seasonal difference in the amount of rain that falls on the average rainy day. In winter the average amount is fairly small, and much of the precipitation comes from cyclonic uplift, which gives widespread, often long-lasting but gentle rain. In summer there is much more convection, with individual storms giving short, intense bursts of rain over a local area.

This information indicates a major characteristic of the weather and climate of North Carolina. We expect precipitation to be abundant and more or less evenly distributed throughout the year, although the character of the rain may vary according to the season. This rainfall is something we tend to take for granted, but it means, for example, that we have continuously flowing rivers while it profoundly influences our natural vegetation and agriculture. We might regard it as normal, but not all parts of the world have similar precipitation regimes.

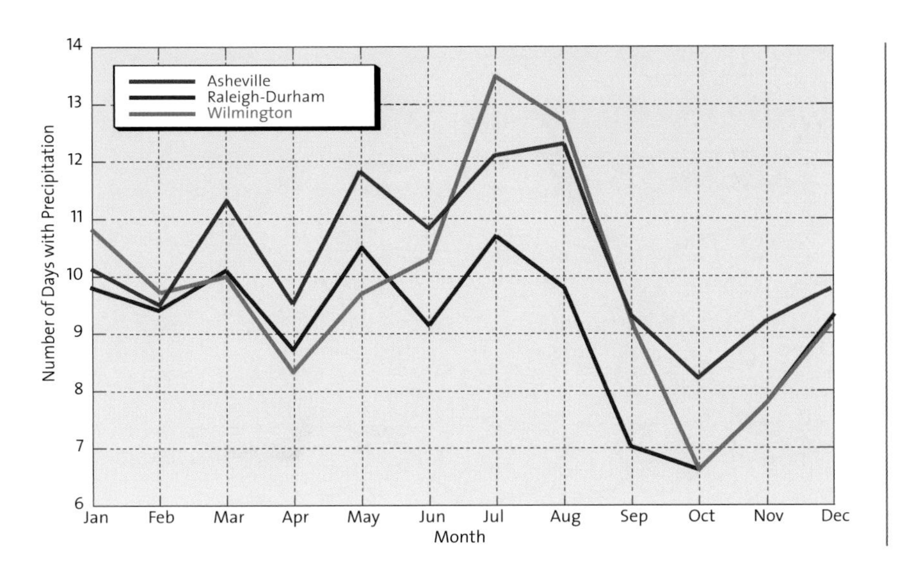

FIGURE 3.12. *Average number of days per month with more than 0.01" of precipitation, for selected cities*

Much of the interior United States has a distinct minimum of precipitation in the winter (see Fig. 3.13). California has a summer minimum. Even the wet months in these areas have only about the same amount of rain that North Carolina receives each month of the year. The desert southwest, of course, has very small amounts throughout the year.

WINTER PRECIPITATION

So far we have not said anything about the type of precipitation that falls. This depends on the temperatures in the air layer between the clouds and the ground. In summer over North Carolina, this layer is usually above freezing, and most of the drops that fall through it start as liquid and arrive at the surface as rain. In winter, however, the situation is more complex. Generally, snowflakes fall from the base of the cloud itself. Most of the time the air layer between the cloud and the ground is above freezing, and the snow melts on the way down, again arriving as rain. If the whole layer is below freezing, snow occurs. Sometimes the middle of the layer is above freezing and the top and bottom are below freezing. Then the snowflakes melt on their way down and are cooled again. If the freezing layer near the ground is fairly thin, the precipitation may arrive as supercooled water—liquid water with a temperature below 32°F. This then freezes on impact, creating ice. If the cold layer is thicker, the raindrops will freeze into ice pellets while they fall, creating the precipitation we know as sleet.

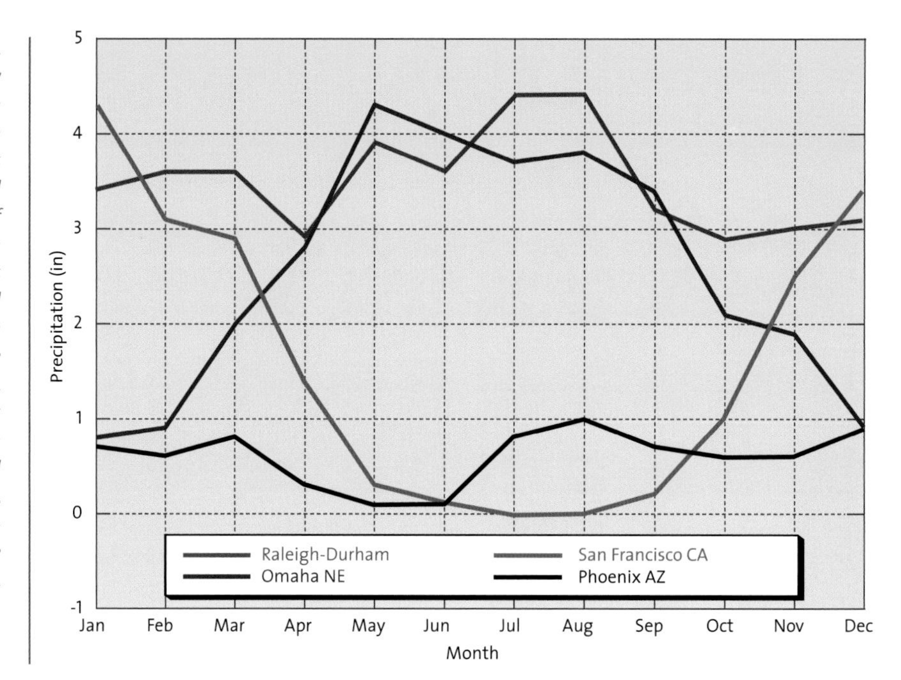

FIGURE 3.13. *Average monthly total precipitation for Raleigh-Durham and selected cities in the United States. Much of the southeastern United States has a slight rainfall maximum in summer, but there is no dry season. Most other areas of the nation have a marked dry season, the Pacific Northwest being the main exception.*

In our state it seems that at least a few times every year we have conditions when the clouds and the ground are below freezing and the middle atmosphere is above freezing. Then we get an ice storm, often with a mix of snow, sleet, freezing rain, and rain. Sometimes the precipitation is scattered in a series of bands across the area, and sometime it all occurs in sequence at a single place. In any event, disruption of transport, to say nothing of damage to vegetation and buildings, is the result. One such event influenced much of the Piedmont from Charlotte to the Triangle on December 4, 2002 (Box 3.1).

Snow is our most common form of frozen precipitation. Amounts vary from place to place, but average annual totals, and the number of days with snow, are greatest in the mountains (see Fig. 3.14). Averages can be misleading, however, since year-to-year variability is very great. The mountains get some snow most years, while the Piedmont seems to get a few inches from storms several years apart. The average annual amount for the Coastal Plain is close to 1.0". For Wilmington it is 2.3". But in December 1989 the city got 13" in a single storm, while 11.7" fell in twenty-four hours in February 1973. Most years the total is 0.

Although there is great year-to-year variability, there does seem to have been a trend in snowfall amounts over the last half-century for which we have good records (see Fig. 3.15). After an almost snow-free decade in the 1950s, the 1960s

BOX 3.1. THE WINTER STORM OF DECEMBER 4–5, 2002

Most storms in winter are rainstorms. Some, especially in the mountains, consist entirely of snow. But some, and usually the most deadly and damaging, are a mix of rain, snow, and ice. That of December 4–5, 2002, was a fine example. For a few days prior to the storm, cold, Arctic air had been sitting over the state. Then, late on December 4, warmer moist air streamed into North Carolina from the west. This warm air rode up over the cold air and created the precipitation. The air also had to rise over the mountains and then descend onto the Piedmont. This airflow pattern led to the creation of bands of precipitation of various types, all arranged roughly parallel to the mountains. The far western mountains had rain; the rest of the mountains and much of the western Piedmont had snow. Moving southeastward the amount of freezing rain increased, reaching a maximum just north of the line between Charlotte and Raleigh. Farther southeast, rain became the main feature. As the whole weather system evolved and moved eastward, one precipitation type replaced another. For most of the Piedmont, for example, the initial freezing rain fostered the accumulation of glaze ice, especially on tree limbs. The weight brought many crashing down, and they usually brought ice-stressed power lines with them. On top of the ice came a layer of snow. The result, typical of an ice storm in modern times, was difficulty of movement over roads; much damage, especially roof damage, to individual properties; and lack of electricity. In this case more than 1 million residents of North Carolina were affected; some were without power for more than a week. Total damage ran into the hundreds of millions of dollars. Few fatalities were directly associated with the storm, but it undoubtedly contributed to many deaths and injuries, because of exposure to cold, stress caused by snow removal, or the increased incidence of traffic accidents.

had numerous snow days. For the rest of the century the number consistently declined. More recent events suggest a renewed increase in the number of snow days.

RAINMAKING

With the abundant rain that normally falls on our state, there is usually little cause to think about rainmaking. But sometimes as summer progresses and drought intensifies, wells dry up, and crops wilt, it seems an attractive proposition. Throughout the world, and probably throughout human history, there have been similar sentiments. Early attempts to make rain were based on crowds at the ground making noise to shake the rain out of the clouds in the air. Cannons and rockets exploding in the clouds were also used in attempts to shake the water out. These methods may cause a slight increase in the amount of water leaving the cloud base, but

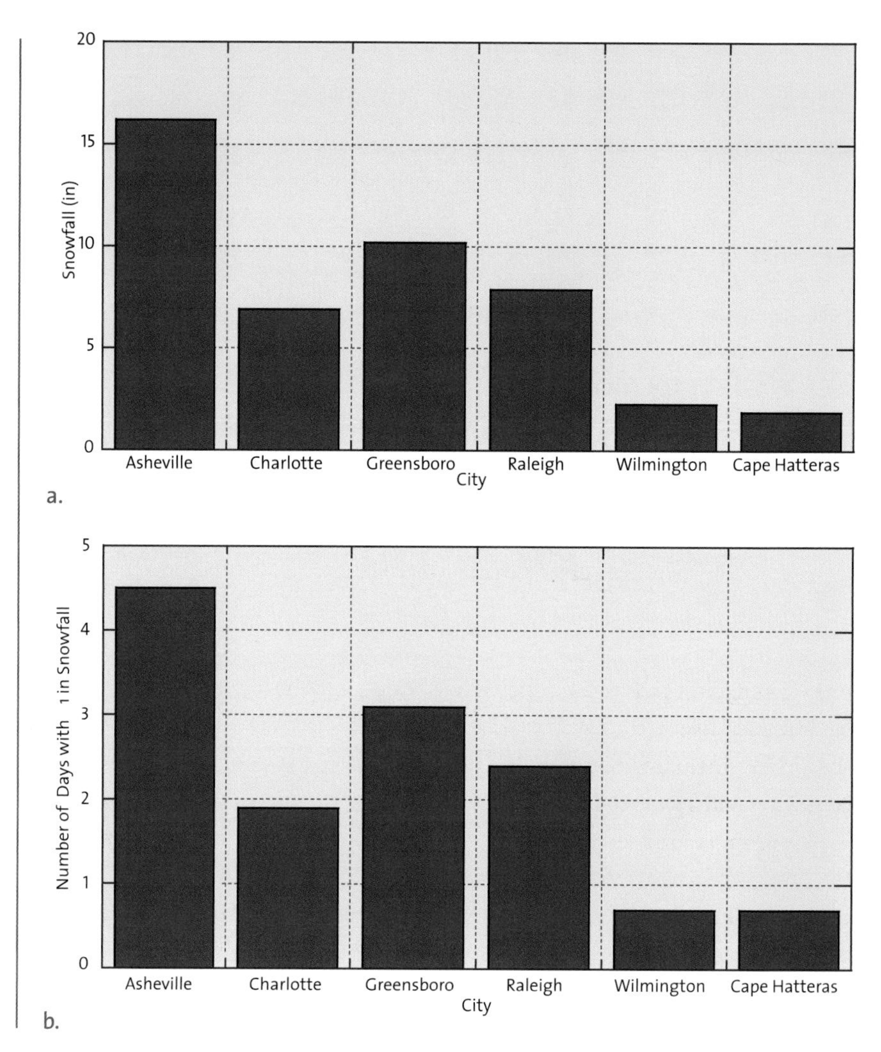

FIGURE 3.14. *Average annual (a) total snowfall and (b) number of days with snowfall greater than 1" for selected stations*

raindrop evaporation on the way to the ground easily removes any excess water, so no precipitation increase results.

More scientifically based experiments undertaken during the middle of the twentieth century soon indicated that the most promising approach was to get the small droplets comprising a cloud to grow into drops big enough to fall. In theory at least, if either silver iodide or dry ice (solid carbon dioxide) crystals were introduced into a cloud, droplets would be stimulated to grow. Hence the term *cloud seeding* was coined. This technique has been widely used, with aircraft flying over the clouds and dropping the seeding crystals. Despite the effort, there is conflicting evidence about how effective the method is. A major difficulty lies in

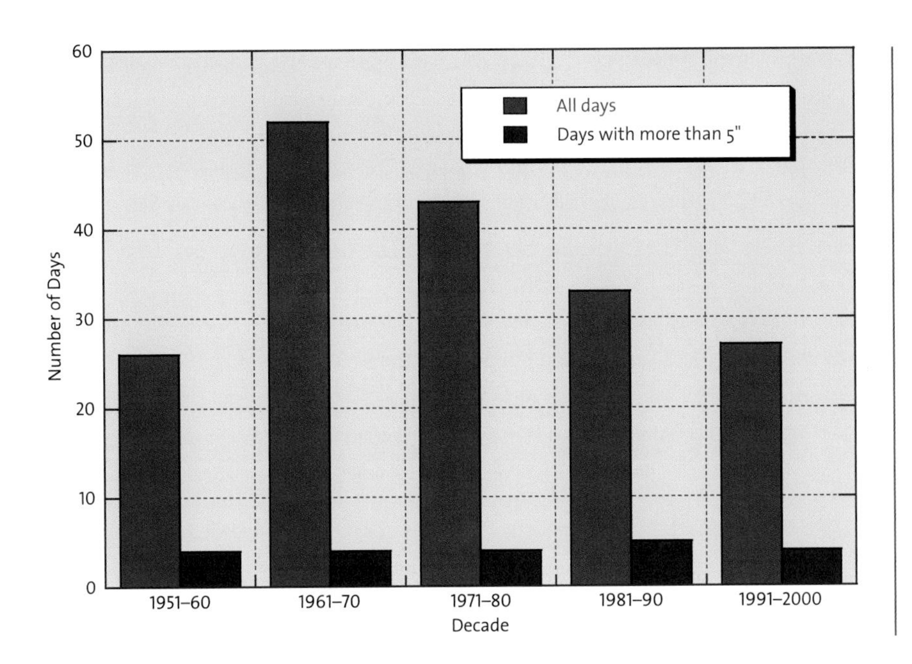

FIGURE 3.15. *Trends in the number of days with snowfall at Raleigh-Durham airport. There was a marked peak in the 1960s, but the frequency has been decreasing since then. Days having more than 5" of snow have always been rare, and the frequency has changed little. Snowfall totals show a pattern similar to that for the total number of snow days.*

determining how much rain a particular cloud might have given even if it had not been seeded. Nevertheless, numerous agriculturalists in many semiarid parts of the world—including many ranchers in states such as Texas and Oklahoma—pay commercial firms to seed the clouds over their land.

The term "cloud seeding" for the process of rainmaking—or more scientifically the process of *precipitation augmentation*—hints that a cloud is needed before we can start. During most of our North Carolina drought periods, especially those midsummer stretches of great concern to agriculture, the whole atmosphere above us seems to be designed to prevent cloud formation. A high pressure region—the Bermuda high discussed in the next chapter—sits over us, forcing air to descend to the ground and slowly drift away and preventing any of the rising motions needed for cloud formation. Thus rainmaking is rarely an option for us.

Water at the Ground

In the final stage in our review of the hydrological cycle, we consider the water that has been returned to the surface of the earth. After arriving at the surface, it may flow away in rivers or seep into the soil. In both cases, the amounts and timing involved are of great practical consequence for crop or garden growth, and for water supply, from individual wells and community reservoirs. Variations in amounts from time to time and place to place can cause problems; sometimes there are shortages, but sometimes there is overabundance. Again, meteorological information is needed to ensure efficient use of the water.

In order to understand and predict the role of weather in water at and below the ground, we need to consider a variety of water flows (see Fig. 3.16). Our major concern is with precipitation and evaporation. The precipitation that arrives and flows across the earth's surface contributes directly to river flow. The water that sinks into the ground, however, follows a much more complex route and may become involved in plant growth and evapotranspiration in the soil, in vertical drainage to storage deep underground, or in horizontal subsurface flow that eventually reaches springs and rivers, where it, too, becomes surface flow.

WATER IN THE SOIL

Water in the soil is vital for vegetation growth. Although water goes into the soil virtually every time it rains, the actual proportion of the rain that enters the soil depends on the type of soil, the slope of the ground, and the vegetation, as well as on the amount and intensity of the rainfall. On a gentle wooded slope in a light rain, almost all water enters the soil, while in a thunderstorm on a steep bare slope most runs off, often leading to flash floods (see Chapter 5).

Water also drains out from the bottom of the soil, a process known as *deep drainage*. The rate of this drainage also depends on the type of soil but is generally less than the rate at which water can be absorbed. So when rain falls, the amount of water in the soil increases; during dry periods, the amount slowly decreases as the water both evaporates and drains away. In addition, there is a maximum amount of water that any soil can hold, depending on its type, depth, and tillage. This is the soil's *field capacity*. If the rain, or any irrigation, puts more water onto a soil at field capacity, most of the excess water will run off while the remainder gives a short-term boost in drainage rates.

Early in the year in North Carolina, when evapotranspiration is low but rainfall is high, the soil tends to be at or near field capacity (Fig. 3.17). As the year progresses, evapotranspiration increases but rainfall amounts stay more or less constant. Then the water in the soil falls below field capacity. In Aurora in both 1994 and 1996 this occurred around the beginning of April. In 1996 there was a fairly short dry spell early in the summer; but starting in June, precipitation picked up, and the late summer was wet. Soils were at field capacity for most of the year. An agriculturalist would not have needed to use much irrigation in this season. But 1994 was a dry year with little summer precipitation. Soil water amounts continued to

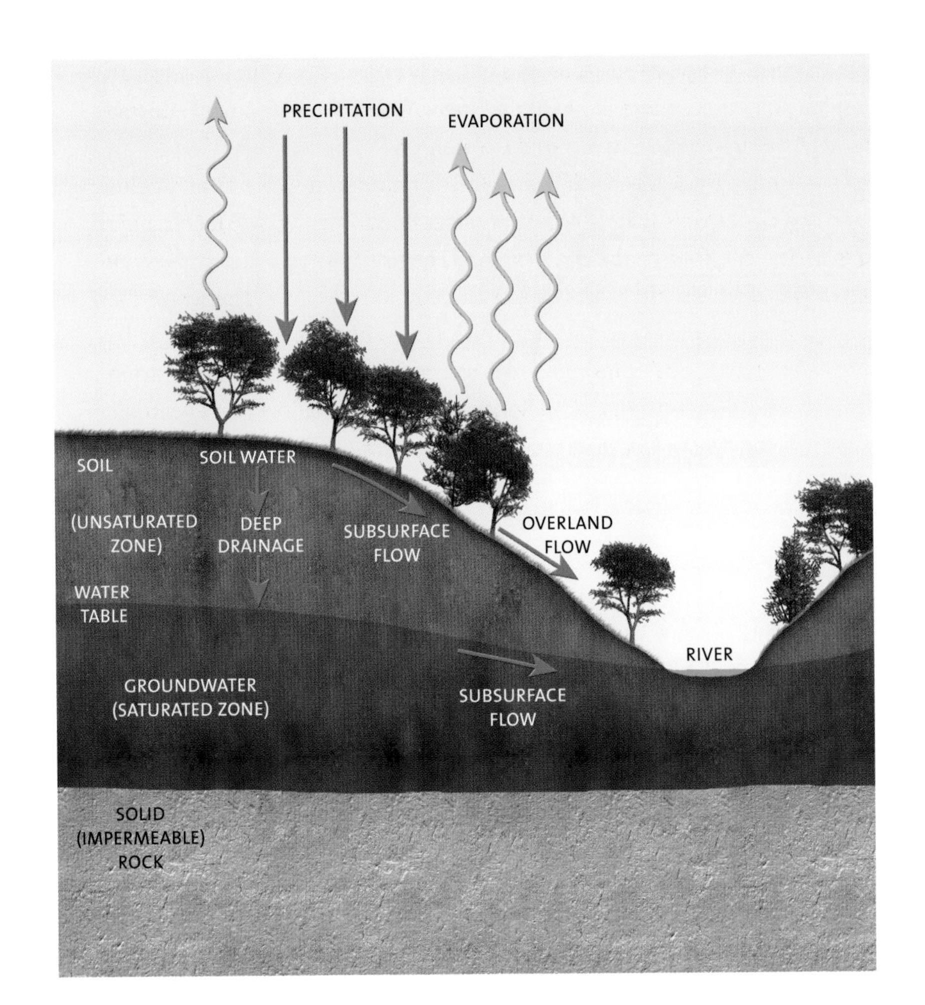

FIGURE 3.16. *A schematic cross section through the upper layers of the earth, showing how water flows and the areas where it collects*

decrease until in the middle of July the soil was almost "empty." It had reached the *wilting point*, where the water content was so small that neither the plants nor the atmosphere could extract more water. Irrigation would clearly have been a benefit in this year. Not until later in the year, when evaporation rates decreased and precipitation returned to more usual amounts, did the soil begin to fill back up.

These two years show, in somewhat extreme form, the characteristic soil water pattern for most of North Carolina. Soils are at field capacity in the winter and begin to dry out in the summer. The amount of drying depends on the summer rainfall amount. During the fall and winter, lowered evaporation rates allow the soil to come back to field capacity.

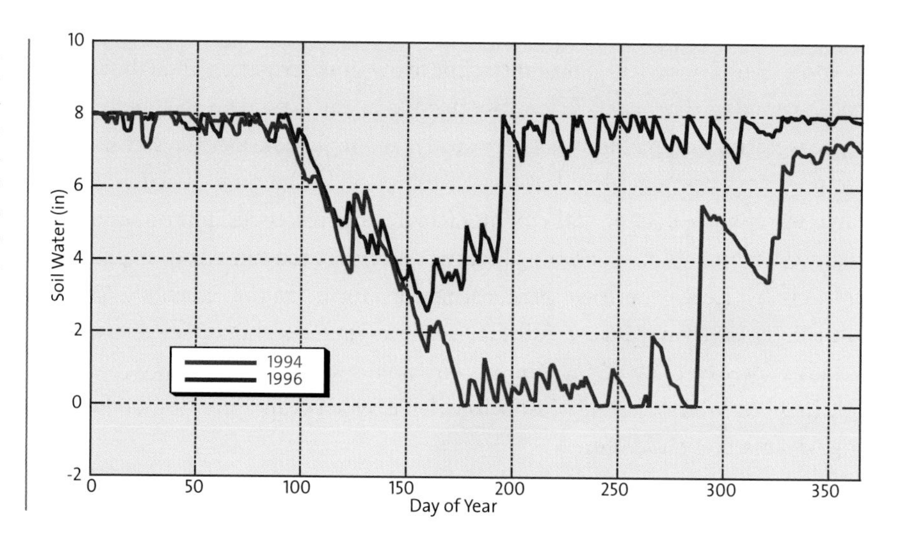

FIGURE 3.17. *Daily soil moisture estimates for Aurora, Beaufort County, in a dry year (1994; 47" total) and a wet year (1996; 61" total). It is assumed that the field capacity for this thick Coastal Plain soil is 8".*

IRRIGATION NEED AND SCHEDULING

Many agricultural crops—and backyard gardens—prefer the soil at field capacity for most of the time. But in most North Carolina summers the soil water falls below field capacity for some of the time. So we have to consider whether it is worthwhile to buy and use irrigation equipment. The answer depends on much more than weather alone. Nevertheless, the meteorologist can help the agriculturalist answer three basic questions:

1. *Will irrigation be a long-term benefit?* It is possible to do an analysis of the kind used in Figure 3.17 for your area for many different years in the past. This will suggest the percentage of years in the future where having irrigation available would help your crop production.
2. *What is the most beneficial strategy for the coming growing season?* This same information about past conditions can be used with seasonal forecasts of precipitation and temperature made several seasons in advance. This helps with decisions, such as those of seed purchase, for the next growing season.
3. *Should I irrigate now?* There is nothing more frustrating—or financially wasteful—than irrigating when the soil is already almost at field capacity or on the day before a soaking rain. Tracking the current soil moisture status using precipitation and evapotranspiration estimates, and knowing the current crop needs, helps us avoid the former. Looking at tomorrow's weather forecast should assure that we avoid the latter.

Most of us are not agriculturalists, but many of us irrigate, although we might call it watering the lawn. We may not need to be involved with the first question, since watering equipment is not very costly. We might consider the second one as we plan next year's garden. But we are likely to ask the third question. Routine lawn watering is a major drain on municipal water resources in the summer, and many governments are increasing their water rates in the summer or imposing use restrictions. Hence there are practical reasons for irrigating efficiently. Tracking the soil moisture may be excessive for a casual gardener, but judicious use of the weather forecast to avoid irrigating before rain is highly practical. So is irrigating late in the evening or perhaps just before dawn, when evaporative loss of the spraying water is at a minimum.

UNDERGROUND WATERS AND WELLS

The water leaving the soil as deep drainage percolates down through a region of broken, weathered rock fragments. Eventually it encounters a bedrock layer, perhaps only a few feet below the surface in the Piedmont and mountains but a thousand feet down on the Coastal Plain, that is impermeable to water. The *groundwater* collects in a saturated layer on top of this impermeable rock. The top of the layer is the water table. The level of the water table does not respond very much to an individual rainstorm, but during a wet spell, water will percolate down and the water table will rise. Conversely, the table will fall during a dry spell. This drop will result partly from loss as water runs away along the top of the sloping water table. This loss will occur more or less continuously. It joins the horizontal flow of the soil water to become the *subsurface runoff*.

We extract groundwater using wells. A well will operate as long as its bottom intake is below the water table. With North Carolina's climate the water table level will fluctuate within the year, being highest at the end of winter and lowest at the end of summer. These fluctuations are usually small, but they are closely related to those for the seasonal change in soil water amount. Much more important is the need for the rainfall to be sufficient to overcome the continuous loss of groundwater to streams through subsurface flow. Any period with low rainfall will almost automatically cause the water table to fall. The extraction of water through the well for human use simply speeds up the rate of fall. During an extended dry period the water table may fall below the bottom of the well. With a dry well, our only options are to seek another source or, if possible, deepen the well. The length of extended dry period needed for a well to go dry depends on the rainfall shortfall, the configuration of the water table, and the amount of water being extracted.

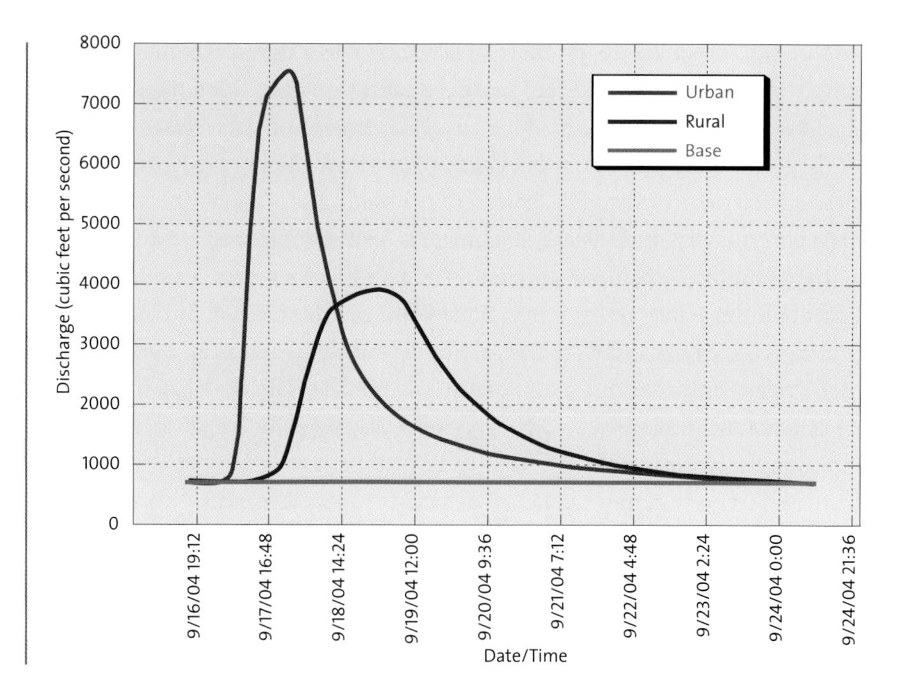

FIGURE 3.18. *A hydrograph showing an almost constant base flow as it would appear in a dry period and the flood flows associated with a similar rainfall over an urban and a rural area. The discharge amounts and the time involved are for illustrative purposes only.*

RIVERS AND RESERVOIRS

The slow subsurface flow, which is ongoing whether or not rain is falling on the ground above, continuously moves down the water table and into a stream. It may come to the surface along the valley side as a spring or, more commonly, feed directly into the stream. This provides the *base flow* of a river (see Fig. 3.18). The base flow is closely associated with the water table level and, like that feature, fluctuates only slowly with time. Base flow is usually at maximum in late winter or early spring and at minimum in the fall.

A river is also fed by the much more rapid direct surface runoff that occurs only during or immediately after rain. This is the *flood flow*. In this context the flood flow refers to the flow in excess of the base flow and has no necessary connection with rivers bursting their banks and causing death and destruction. Since we expect more rainwater to run off the wet soils of winter than the drier ground of summer, winter rains are more likely than summer rains to give flood flows. The seasonal difference is made more complicated, however, by the nature of the rainfall. In summer we have many intense thunderstorms with rain falling much faster than any soil, however dry, can handle. Much of this rain runs off to become flood flow. The good "soaking rains" that replenish the soil are usually associated with convergence. We do get them in summer, when they affect the river flows in only minor ways, but they are most common in winter, when the soil is often already nearly full and they produce broad flood peaks on the rivers.

The amount of water available to us through stream flow depends on almost the same meteorological factors that affect the groundwater. This may be unfortunate, since it means that at the same time that our wells are running dry our rivers are probably running low. One solution is to dam the river and create a reservoir that stores the water so that it is available during a dry spell. If dam construction is to be considered, weather information is needed so we can determine the size, design, and most efficient way to operate the dam. Indeed, the types of information needed are very similar to those discussed for irrigation scheduling. Past weather observations are used to determine how much water a dam of a particular size would trap and how much the reservoir level is likely to fluctuate in response to dry and wet spells. The calculation methods used—much like those for chance of frost occurrence (see Table 2.3)—must be based on probabilities using past information, since we are a long way from being able to give forecasts for the multidecade life of a dam. But we can give the seasonal and daily forecasts needed to help with reservoir operation, thus ensuring that the water level becomes neither too high and overtops the dam nor too low and causes the water supply to dry up.

chapter 4

Our Common Daily Weather

In this chapter we look at our day-to-day weather—not the spectacular events, but those that are by far the most frequent. First we look at the winds, which, in many ways, we can envision as bringing in this daily weather. Then, in order to understand and explain our own weather, we have to look briefly at the weather of the whole world. This leads to a description of the daily weather sequences we have in North Carolina.

Winds in North Carolina

One of the most useful ways to make a quick, short-term weather forecast is to look at the sky in the direction from which the wind is blowing. Cloud patterns there usually give a good indication of what is coming our way. However, close observation of the movement of clouds will soon indicate that they do not always travel in exactly the same direction as the wind we experience at the surface, and clouds at different levels may move in different directions. While the amount of these differences is sometimes very important for detailed forecasting, most of the time the clouds and the air at the surface move in the same general direction. There is also a difference in wind speed with height. High clouds almost always move faster than low ones.

The difference between wind conditions near the earth's surface and those aloft is primarily the result of friction at the surface. The amount of friction, and thus the amount that the surface air is slowed and its direction changed, is determined by the *surface roughness*. This depends on the local topography and the number, size, and distribution of obstacles—including grass, trees, and buildings—to the airflow. The rougher the surface, the greater the slowing and twisting of the wind.

Roughness varies from spot to spot, so the surface airflow pattern tends to be very complex. This also means that we cannot make statewide maps of surface winds. Instead, we have to look at individual stations.

THE MOST COMMON WIND DIRECTIONS

By far the most common wind direction for our state is southwesterly. (Remember: we always label the wind by the direction *from* which it is blowing). This southwesterly airflow, which usually brings warm, moist conditions, dominates in all seasons. In spring and summer the southwest is virtually the only direction from which winds commonly blow. Winter, however, has frequent northerly winds, with directions extending from northwest to northeast. This reflects the frequent outbreaks of cold polar air common at this time. Northeasterly airflow is also frequent in fall, largely as a result of the changes in the global air circulation. But it also reminds us that this is the season for hurricanes and nor'easters. Winds from the east are rare at any time of year.

WIND SPEED

There are seasonal variations in wind speed as well as direction (see Fig. 4.1). Late winter and early spring give the fastest average wind; late summer and early fall, the slowest. Asheville has the greatest seasonal difference. This is probably because of the station's sheltered location in a basin in the hills. Since we have no other reliable long-term observations in the mountains, we tend to accept the Asheville values as typical of the whole region. Greensboro and Wilmington, on the other hand, are very typical of the Piedmont and Coastal Plain, respectively. They have almost identical patterns, with Coastal Plain winds being faster by about 1 mile per hour (mph) throughout the year.

These values are average wind speeds throughout the day. There is almost always less wind at night than during the daylight. During the day, solar heating, both locally and on a continental scale, causes hot spots and cool regions, which in turn help to generate wind. At night, temperatures are much more even, so there is less wind.

The change in wind speed from place to place is most noticeable along the coast, where there is a rapid increase in wind speed as we get near the ocean. The lower friction of the smooth water surface does not slow the air as much as the rougher land. The station at Wilmington airport is only a few miles inland, but it is positioned well behind the rough dunes and is much less windy than Cape

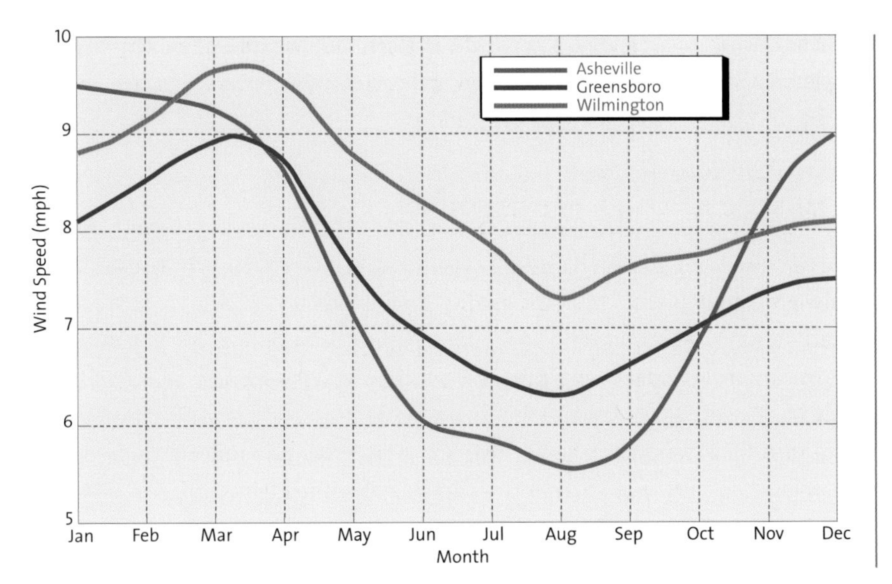

FIGURE 4.1. *Average wind speed for each month for selected stations. The exact values depend on the exposure of a particular station, but for all there is a clear minimum in summer.*

Hatteras, closer to the shore. Indeed, because of friction, on almost any day, kite flying is possible on the beach but not a few hundred yards inland. Similarly, in the mountains the exposed peaks tend to have much higher wind speeds than do the sheltered valleys. Indeed, it is in these valleys where calm conditions are most likely. Averaged through the state, calm conditions occur about 10 percent of the time. They are most frequent in late summer and fall and least in spring.

So far we have been considering monthly average wind speed. This is determined from observations, every hour on the hour, of the two-minute average wind speed. A two-minute average smooths out the gusts and lulls in the wind, giving a realistic estimate of the overall speed. Our observing instruments also commonly record the *peak gust speed*—the highest five-second average detected within the hour. This is probably not the actual maximum wind speed, since very strong gusts often last much less than a second. But it is the closest that the rugged instruments needed for routine, continuous operations can come to a true maximum.

The most spectacular winds, of course, are associated with hurricanes. Even when there is no hurricane, it is not uncommon to have an event when the two-minute average wind exceeds 25 mph, with peak gusts at nearly double this speed. These gusts can create property damage and be dangerous for pedestrians. Many people have difficulty staying upright in the variable gusts and lulls associated with a 25-mph wind, while a 30-mph wind is physically dangerous for some people, mainly the very young or the very old. In our summer these winds are usually localized and associated with downdrafts from thunderstorms. In winter they are associated with moving weather systems and can cover a broader area.

KITTY HAWK WINDS, 1903

One of North Carolina's major claims to fame—First in Flight—is due largely to the wind conditions on the Outer Banks and the dedicated observers who recorded the wind speeds.

When Orville and Wilbur Wright were looking for a site for their first field experiments in 1900, they needed a place where the winds regularly blew faster than 15 mph to provide enough lift for their aircraft's wings, some gentle hills to allow glider launching, a sandy or similar surface for soft landings, and a remote location so they could concentrate on their work without the distraction of public or media attention. In late 1899 they wrote to the U.S. Weather Bureau asking about wind velocities around Chicago in the fall. The reply suggested that nowhere in the Chicago area would meet their needs, but a copy of the September *Monthly Weather Review* was included. This contained a table of monthly average wind velocities from approximately 150 stations nationwide (extracts from this table appear in Table 4.1). The values were not encouraging. Few stations had speeds greater than 15 mph. Even if 12 mph was used as the wind speed threshold, there were still few candidates. Hatteras and Kitty Hawk seemed the best of a less-than-ideal bunch.

On August 3, 1900, Wilbur wrote to the Weather Bureau office at Kitty Hawk asking about conditions. A quick reply from the sole employee, Joseph J. Dosher, said there was a beach a mile wide without trees or other obstructions. Winds in September and October blew from the north or northeast. Dosher also passed the letter to William J. Tate, postmaster, notary, and Currituck County commissioner. Tate sent a more detailed and equally favorable description of the area and offered a warm welcome. The Wrights arrived in Kitty Hawk that September and spent the fall testing glider configurations. A return visit in 1901 was ruined by bad weather; but the next fall was more fruitful, and they gathered the glider performance information needed to develop a powered machine.

In 1903 the brothers arrived at Kitty Hawk on September 26. They immediately performed some tests with the 1902 glider. Many days of bad weather delayed progress, but by November 5 a powered machine was ready for ground testing. Refinements, adjustments, and propeller problems delayed progress. But by December 12 the machine was ready, and on December 14 the wind picked up. A launch was made, but the craft stalled as it went down the sloping track of Kill Devil

TABLE 4.1. Monthly Average Wind Speeds (mph), August–November 1899, for Stations in the *Monthly Weather Review* Having Wind Speeds of 12 mph or Greater during at Least Three of the Four Months

	AUG	SEPT	OCT	NOV
Amarillo, Tex.	13.2	14.6	16.4	9.9
Block Island, R.I.	10.6	15.4	15.6	16.2
Cape Henry, Va.	14.2	12.0	13.3	12.4
Chicago, Ill.	13.8	16.9	17.6	17.1
Hatteras, N.C.	14.0	11.2	13.0	12.5
Kitty Hawk, N.C.	13.9	13.4	16.3	14.3
Mt. Tamalpais, Calif.	16.5	17.1	18.2	16.7
Sandy Hook, N.J.	12.6	17.1	15.1	18.2
Sioux City, Ia.	12.0	12.8	12.5	n/a

Note: Other stations along the southern Atlantic coast—Norfolk, Va.; Wilmington, N.C.; Charleston, S.C.; Savannah, Ga.; and Jacksonville, Fla.—had average wind speeds less than 12 mph.

Hill. Repairs were needed. These and poor weather delayed the next trial until December 17. By then the weather was frigid, but the wind was over 20 mph. Four flights—the last of 852 feet in 59 seconds—and one famous photograph were made that day. The experiment ended when a wind gust lifted one wing, overturned the aircraft, and destroyed it. But history had been made.

The *Monthly Weather Review* for December 1903 included an article headed "Meteorology and the Art of Flying." The concluding sentence stated, "Their success is undoubtedly due in great part to the preliminary careful study of the winds, for this reason, although machinery is essential, yet we consider that meteorology also has played an important part in their work." We also consider that climate played an important part in ensuring North Carolina's place in this particular historic event.

North Carolina's Place in the World of Weather

The winds are not simply a part of the weather we feel every day. They are also vital because they largely control the direction from which weather arrives over our state, while through the wind the atmosphere above our heads is connected to that of all other places on earth. Our weather may have its origin in places far removed from our own location. The winds themselves are a response to changes

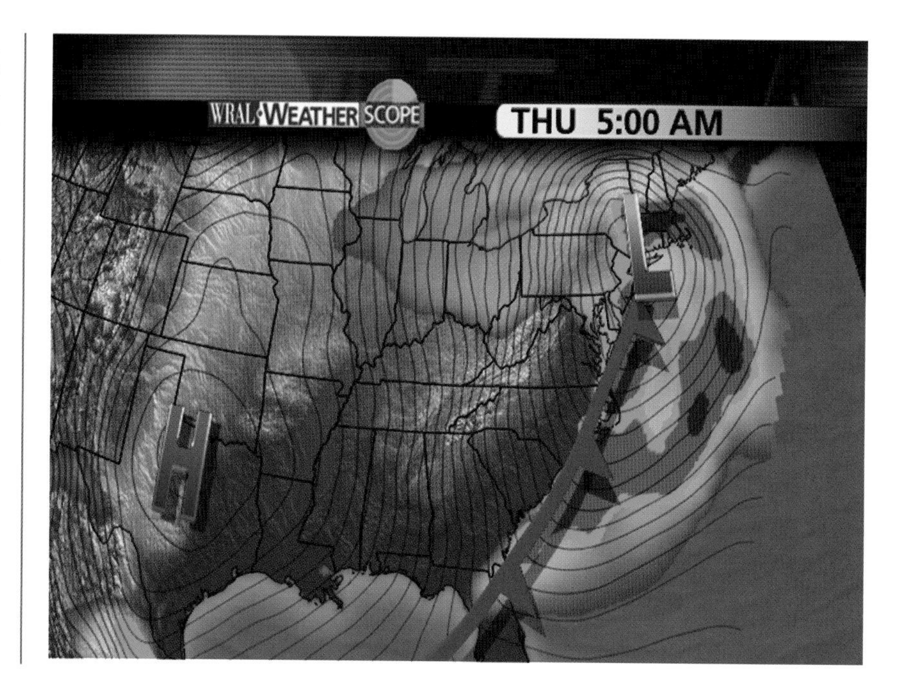

FIGURE 4.2. *A weather map of the type commonly seen during TV weathercasts. The dark lines are the isobars joining places with equal pressure, centers of high (H) and low (L) pressure are indicated, with a (cold) front extending south from the low pressure center. The green regions indicate clouds. A map of the situation twenty-four hours later is shown in Figure 7.5. (image courtesy of WRAL-TV)*

in atmospheric pressure, and we often, particularly in weather forecasting, spend much time considering pressure and pressure changes even when we are mainly interested in the winds that result.

The air flows—the wind blows—in response to the horizontal differences in pressure. One of the major pieces of information any meteorologist needs is a map showing this distribution (see Fig. 4.2). Commonly we need to know the surface distribution, indicated on the map by lines, called *isobars*, joining places with equal pressure. Pressure forces air to move from regions of high pressure to regions of low pressure. However, the rotation of the earth causes a deflection in the air, and the actual flow is almost parallel to the isobars. In our Northern Hemisphere the deflection causes air to blow so that, if we stand with our back to the wind, low pressure is on our left. On the day shown in Figure 4.2, air is coming into North Carolina from the north, and on the coast it swings around to leave from the west.

Figure 4.2 shows a fairly common situation, having centers of low and high pressure with more or less circular isobars around the center. Air flows counterclockwise around a low pressure system and clockwise around a high. The closer together the isobars are, the faster the wind will blow. A high pressure area and its clockwise flow is called an *anticyclone* (see Fig. 4.3). The opposite, the low with counterclockwise flow, has several names, depending on the size and intensity of

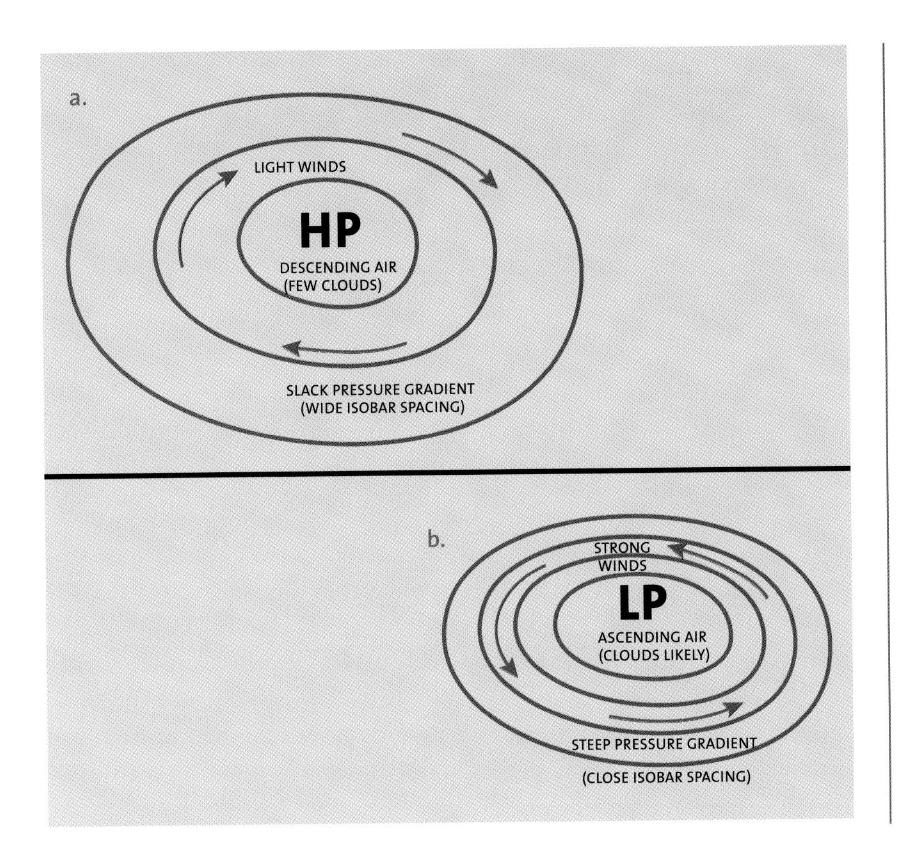

FIGURE 4.3. *Schematic diagram of (a) anticyclones and (b) cyclones. The former have a high pressure center, descending air, and winds blowing in a clockwise direction. Usually the pressure gradient is slack, winds are light, and conditions are often cloudless. In cyclones there is a low pressure center and counterclockwise, ascending winds. Winds may be strong, and clouds are often associated with cyclones.*

the feature. Hurricanes, tornadoes, depressions, and cyclones are all names we shall encounter eventually. For now we can give them the general name of *cyclonic storms*.

GLOBAL WINDS AND WEATHER

Pressure patterns such as those of Figure 4.2 are part of the global system of pressure. The high pressure over Texas is part of a persistent feature of the planetary atmosphere, the global belt of high pressure at about 30°N. There is another at about 30°S. Between them is a low pressure area encircling the globe, often called the equatorial trough. There is a much more diffuse region of low pressure as a belt around 60°, and there is high pressure at the poles (see Fig. 4.4).

These pressure belts persist year-round, and North Carolina's generally westerly winds are a consequence of the difference between the high pressure belt to our south and the low pressure to the north. The belts vary in position and

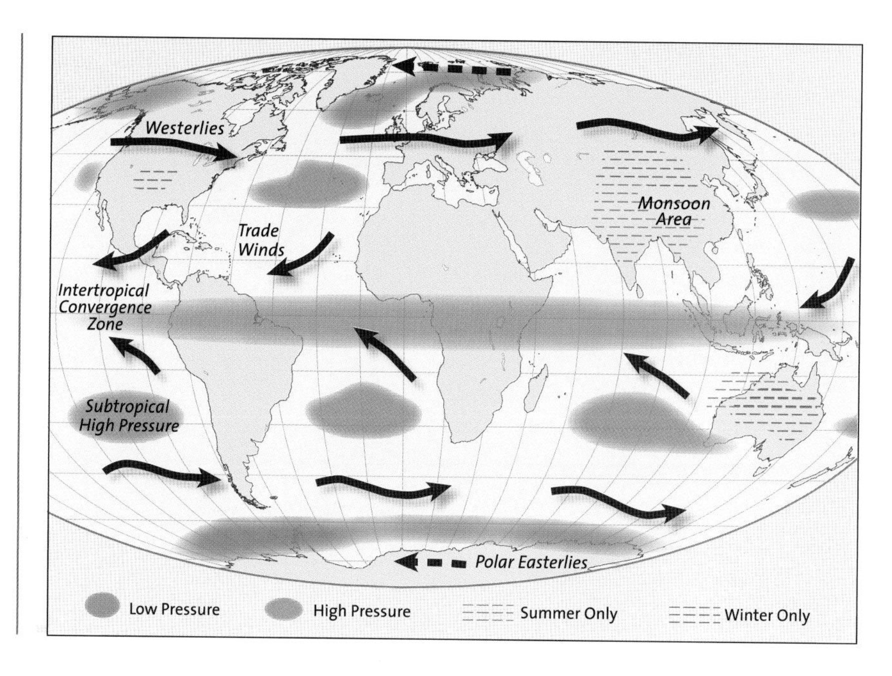

FIGURE 4.4. *Major pressure patterns and wind systems of the earth. Taken together, these are known as the general circulation of the atmosphere.*

intensity. In particular, the low pressure belt around 60°N is much more marked in winter than in summer, a change responsible for the increase in the strength of our winter winds. However, superimposed on this general pattern are features arising from the effects of continentality. Cold areas, continental interiors in winter, lead to high pressure, while warm areas, the summer continental interiors, have relatively low pressure. Over Asia these differences give rise to the monsoon climate, but for us they primarily influence the strength, direction, and temperature of the westerly airflow arriving over the state.

The links between temperature, pressure, and winds are global in scope, and what affects one place often has an impact on places far away. This global system of linkages, known as the *general circulation of the atmosphere*, is becoming especially important as we extend our forecasting further into the future. The experimental forecasts for the number of storms in the Atlantic basin in an upcoming hurricane season, for example, require understanding of linkages involving most of the tropics and much of the Pacific basin.

THE ROLE OF THE OCEANS IN WEATHER AND CLIMATE

There is a *general circulation of the oceans*, a set of currents near the ocean surface and return flows at depth, that complements the atmospheric circulation and influences our weather and climate. Two surface currents influence our weather directly: the Gulf Stream and the Labrador Current. The warm waters of the northward-moving Gulf Stream more or less parallel our southern coast as far north as Cape Hatteras. Beyond that they move offshore and start a journey across the northern Atlantic Ocean. At almost any time in the year the warm offshore waters tend to increase cloudiness in the air passing over them. This seems most marked in winter, when the land-sea temperature contrasts are greatest. The northeast part of the state is also affected by the southern extension of the cold Labrador Current. This at times may penetrate as far south as Cape Hatteras. This current tends to decrease cloudiness. However, because our prevailing winds are from the west and thus blowing offshore, the inland penetration of oceanic effects is usually small. At times complex interactions between the two ocean currents and the land surface play a major role in creating nor'easters, cyclonic storms we will consider at the end of this chapter. Near the coast itself, sea breezes are created; but these are rather local features, so we look at them as part of the overall coastal climate in Chapter 6. Further, in Chapter 7, we will discuss links between the oceanic and atmospheric circulations that are helping us to develop long-range climate forecasts, particularly in reference to the El Niño feature of the Pacific Ocean.

TROPICAL AND POLAR CLIMATES

We need to say only a few words about tropical and polar climates, since it is unlikely that they will have, now or in the near future, a major impact on our own climate. This is not to say that air that comes from the polar or tropical regions does not affect us. As we shall see in the next section (see Fig. 4.5), tropical air and polar air have a major influence on our day-to-day weather.

Meteorologists loosely define the tropics as the area between 30°N and 30°S. Our state, with a southern border around 34°N, is not too far away. However, our current climate, like that of most of the southeastern United States, is certainly not tropical. We have distinct hot and cold seasons and rain throughout the year. In contrast, Florida south from about 29°N (the Tampa-Orlando area) has a climate typical of the subtropics: no real cold season and precipitation primarily in the summer months. Even though the tropical climates are not far away as the crow flies, they are a great distance away meteorologically. Major shifts in the general circulation would be needed before the tropics expand much beyond their current position. North Carolina will probably have climate changes in the future, and we might get wetter or drier, or warmer or colder; but it seems unlikely that we will lose our distinct seasons, and it is very unlikely that we will have the same characteristics as Miami—or even Orlando—does now.

At present we are many miles from polar conditions. It certainly gets colder in both winter and summer as you head north in North America, but you need to be in northern Canada around 60°N before you reach a truly polar climate. However, during a time that ended about 13,000 years ago, North Carolina was much colder than at present, perhaps approaching polar conditions. This was the most recent great Ice Age. On our highest mountain peaks it tended toward a literal ice age, with snow lying most of the year, although there were almost certainly no permanent glaciers this far south. Nevertheless, the vegetation of the area remembers these cold conditions, and we still have in many of our highest places species that we now associate with latitudes much farther north.

Our Midlatitude Weather Patterns

Lying between the tropical and polar climatic regions and covering the area between 30°N and 60°N are the midlatitudes, the region of our own weather and climate. In the simplest terms, our weather consists of a series of events moving across the state generally from west to east, each lasting a few days and each, in turn, being replaced by a new event. The most common events are of two main types: (1) cloudy, usually rainy, and windy storms, often associated with weather fronts and variable temperatures, and (2) calm periods of sunny, often almost cloudless weather where temperatures change little from day to day and where the wind is light and variable. To give these their meteorological names, the storms are *wave cyclones*, and the calms represent our *air mass* weather. In the times of transition between them, when neither is dominant, the general midlatitude westerly airstream, simply called the westerlies, is the major influence. The whole scheme is summarized in Figure 4.5.

There are seasonal variations in the sequence of these events. In summer an air mass giving hot, humid, hazy, cloudless, and nearly calm weather tends to dominate for much of the time (see Fig. 4.5a). This may be interrupted by the passage of rather weak cyclonic storms that bring cooler, cloudy conditions, perhaps with a little rain. In winter, in contrast, the cyclonic storms are much more vigorous and give more rain or snow (see Fig. 4.5b). They tend to move through the state fairly quickly and last only a couple of days over a particular place. Between these storm passages we get the air mass weather, but in winter we get two very different types. One is akin to that of summer, but it is warm rather than hot, though still very cloudy and humid. The other gives the dry, cloudless, and very cold events we commonly refer to as cold Canadian air. Indeed, this cold air does often come from Canada and is moved into our area by the action of the westerlies.

This brief overview emphasizes that the westerlies, air masses, and wave cyclones are virtually inseparable components of our weather. But we have to look at them separately in order to understand more fully how they act and the weather they bring.

The westerlies are a broad, eastward-flowing air current covering the latitude band roughly between 30°N and 60°N. They have three components: (1) *Rossby waves*, (2) *jet streams*, and (3) the *polar front*.

THE ROSSBY WAVES

Within the broad flow of the westerlies there is often a fairly narrow but well-defined and well-organized current of air, rather like the meandering main current of one of our larger rivers. This current and its meanderings are termed Rossby waves after the Swedish meteorologist who first analyzed them (although they are often, and more ambiguously, called either the planetary or the long waves). The positions slowly change from day to day, but Figure 4.5 shows typical positions for summer and winter. These waves are always present, although their position, intensity, and organization can very greatly from day to day.

Figure 4.5 shows only the North American section of the Rossby waves. However, these waves flow around the whole planet, completely encircling the pool of cold air that occurs over the North Pole. In summer this cold air pool at the North Pole is fairly small and rather warm, and the Rossby waves in eastern North America are likely to flow eastward between the Great Lakes and the Hudson Bay. North Carolina is in the warmer air to the south, and only occasionally do short-term changes in the flow patterns affect us directly. But in winter the polar cold air pool gets colder and expands, pushing the Rossby waves southward. They often flow south of the Great Lakes. Since they are much closer to us, we are much more likely to be influenced by variations in the flow pattern.

JET STREAMS

A jet stream is a thin, flat ribbon of fast-moving air embedded in a generally slow-moving airstream. There are several types of jet streams that occur in the atmosphere. Some are close to the surface, cover a short distance, and last for a

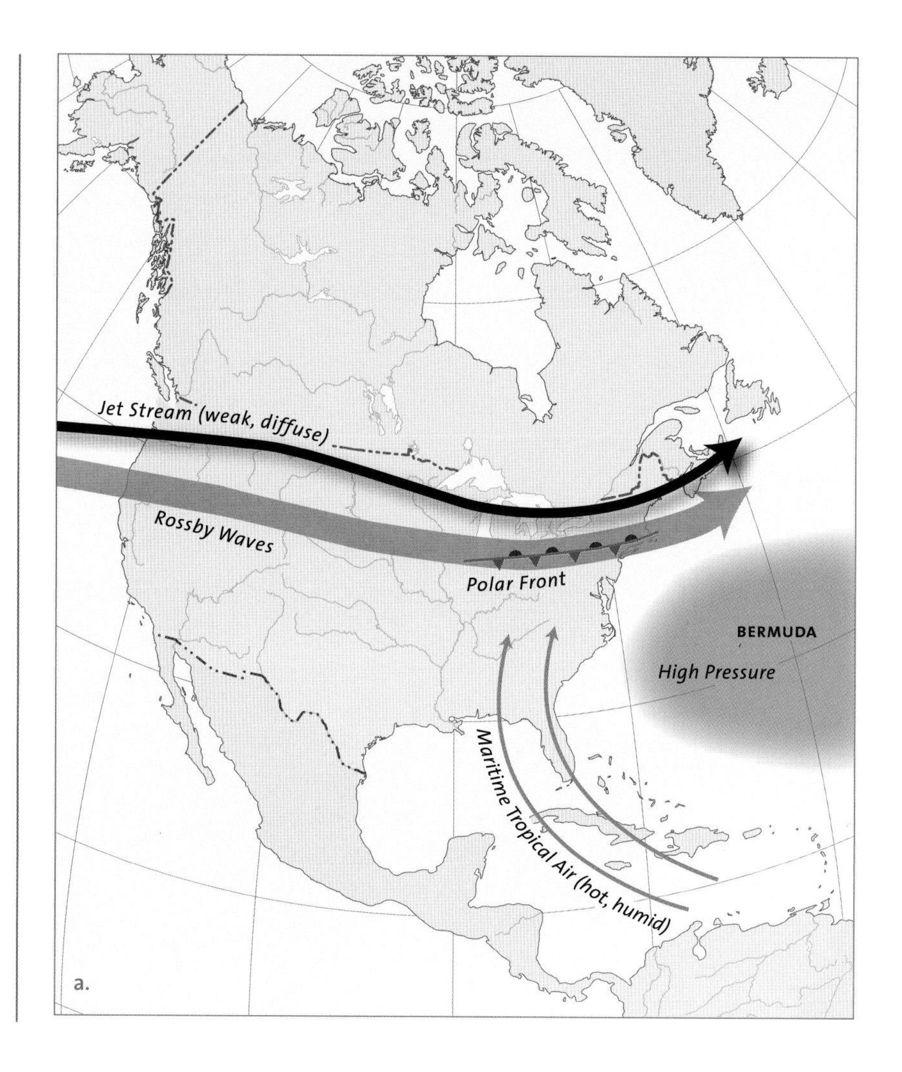

FIGURE 4.5. *Schematic map of typical patterns of features creating the weather most likely to influence North Carolina in (a) summer and (b) winter. In both seasons the jet stream, polar front, wave cyclones, and air masses create the weather, but their positions and intensities vary.*

few hours. Others are much higher, longer, and more persistent and are usually more important for our day-to-day weather. Most of the time we are involved with just one, the polar front jet stream, usually simply called "the jet stream" or even just "the jet." This tends to lie more or less above, and slightly to the north of, the Rossby waves. The two interact with each other to cause changes in either or both of them. But the jet stream is a much better defined and more readily detectable feature than the waves themselves. As such, it has come to play a vital role in the display of current weather conditions and in forecasting weather. Although the stream is not always present or may be split into several different jets (Fig. 4.6), few forecasts can be given without reference to it.

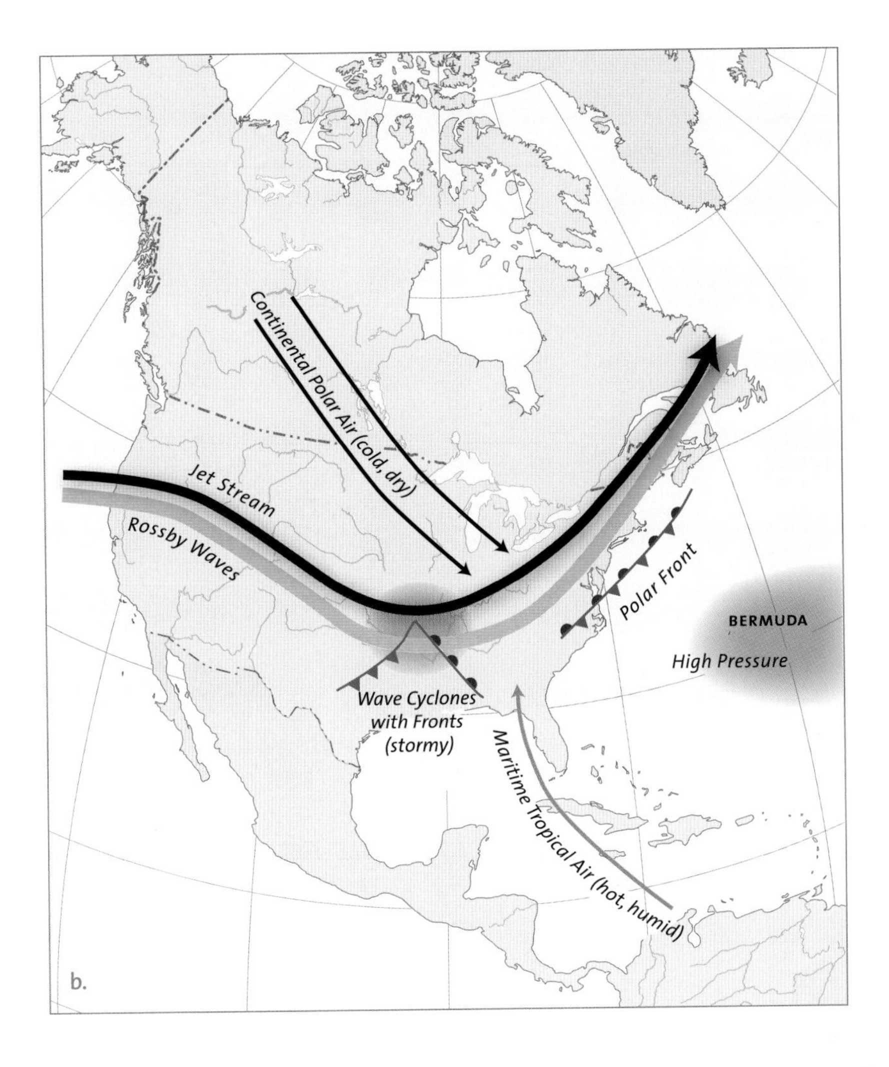

Forecasting the position of the jet stream is of practical importance for commercial and military aviation. Eastbound flights make use of the jet when possible, since it provides a push for the aircraft, thus increasing airspeed and allowing more economical use of fuel. Westbound flights need to avoid the jet. Its position therefore influences route choice on a day-to-day basis. Seasonally the likely position of the jet stream must be taken into account by commercial airlines as they develop schedules. We see the effect on the transatlantic flights from Charlotte and Raleigh-Durham airports. It is much faster to get to Europe than to return, and the difference is greater in winter than in summer. For transcontinental flights, the shorter distances involved and the flight safety rules

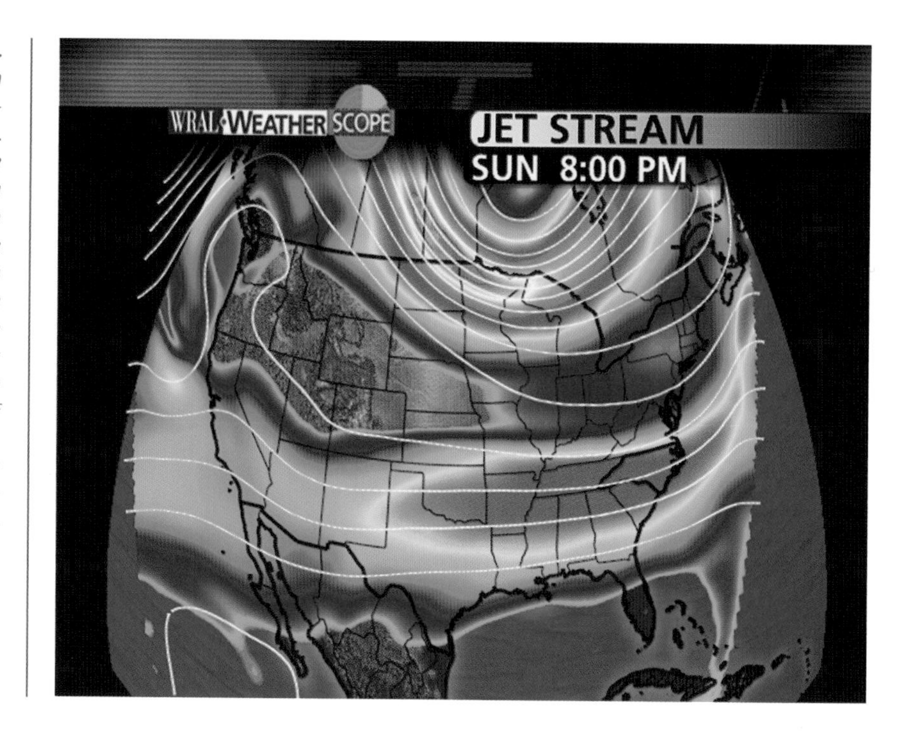

FIGURE 4.6. *Polar front jet stream seen during a summer TV weathercast. The high-speed core of the jet is sweeping in an arc from Canada along the Minnesota-Canada border toward the Great Lakes before losing intensity as it returns to Canada well to the north of New England. A second, much less intense jet gets organized along the Texas-Oklahoma border and flows directly above North Carolina and over the Atlantic. (image courtesy of WRAL-TV)*

imposed in our crowded skies mean that as passengers we rarely notice the effect of wind on airline schedules. Individual flights, however, can be severely affected.

Meteorologically the jet tends to indicate the flow for the whole depth of the lower atmosphere, suggesting the conditions that might be transported into an area. So Figure 4.6 gives a generalized but very pertinent suggestion that North Carolina is likely to be influenced by air from the west. Further, jet streams have a well-defined structure, and we can think of them as tubes with an entrance and an exit. At the entrance, particularly on its south side, air is sucked upward from the surface, producing low pressure there and fostering the development of wave cyclones. At the exit the air commonly descends, and any surface storms are suppressed. So we might suggest that Figure 4.6 indicates that a storm is likely to form in the Texas area and track toward us in the near future.

THE POLAR FRONT

A weather *front* is a region of rapid horizontal temperature change caused when two airstreams or air masses with different temperatures come together. This convergence leads to vertical motions, clouds, and rain and is a vital weather producer. Fronts come in a variety of sizes and configurations; many are associated with wave cyclones. Here we are only concerned with the biggest and most persistent of them all, the polar front. This is the front that separates the tropical from the polar air. The polar front is a deep feature extending many thousands of feet into the air and commonly approaching a jet stream near the tropopause. It may also extend a thousand miles or more horizontally but be only a few tens of miles wide.

The earth's atmospheric circulation is such that there tends to be a broad area of more or less uniform temperature in the tropical areas and another broad area of uniform, but much colder, temperature in the polar regions. We shall consider these features, air masses, in the next section. The polar front separates the polar and tropical air masses. But it is also a feature that is an integral part of the westerlies. Its position varies from day to day in response to changes in the positions of the jet stream and the Rossby waves. As it moves, it pushes the air masses in front of it, causing weather changes in the areas over which it passes. Sometimes the air masses themselves may move in response to changes outside the westerlies, and then they can be responsible for moving the polar front, the jet streams, and the Rossby waves into a new pattern. So all of the features are completely interconnected. For North Carolina in summer, even allowing for variability, the polar front remains to our north most of the time, and we are in the tropical air. Even if the polar front swings south and we are invaded by polar air, in summer this is not too cold, and we do not experience a marked temperature change. In winter, however, the front frequently swings backward and forward over us. So we have warm humid days followed by a rapid temperature fall and a period of cold, dry weather, which in turn is replaced by more warm weather as the polar front moves back north over us.

LINKING UPPER AIR CIRCULATIONS TO SURFACE WEATHER

Much of the time midlatitude weather forecasting begins with a consideration of conditions some distance above the surface. We do this because, first, the clouds form there, and the winds at those levels move the clouds around. Second, the surface and the upper levels largely act together, so that knowing what is happening at one level means that we have a good idea about the other. That leads to a third, more practical reason: it is much easier to determine the pertinent conditions aloft than at the surface. Each spot on the earth's surface simply has too many unique local features influencing our observations for us to sort them all out. Nevertheless, we are primarily concerned with the weather as it affects us at or near the earth's surface.

There are two key links between the upper air and the surface. The first is connected with the Rossby waves. Air blows in a clockwise motion as it swings northward around a high pressure ridge, and then it blows counterclockwise around the southern bend of a trough. Wind blowing clockwise tends to move fast in comparison with wind moving counterclockwise. So as the air moves southeast after a northern bend, it is decelerating and "piling up" (see Fig. 4.7). The piling up, in fact, means the air is moving downward toward the surface. This gives high pressure at the surface, which usually is associated with calm, clear weather. On the other hand, as air moves northeastward after a southern bend, it is accelerating, sucking air up from lower levels, creating low pressure at the surface, and allowing the development of wave cyclones and stormy weather. Jet streams provide the second link between the surface and upper air. As we indicated for Figure 4.6, rising motions are associated with the entrance to a jet stream, sinking ones with the exit. In practice the Rossby waves and jet streams commonly work in tandem to provide vertical motions and create the wave cyclones that are a major distinct feature of our daily weather.

Air Masses: Calms between Storms

Air masses are blocks of air covering about a million square miles, with horizontally uniform temperature and humidity. They form in *source regions*, usually in eddies on the edges of the westerlies, and move from their source regions when the westerly flow patterns change. Although there are several types of air masses, we are particularly influenced by two: one has its source over the tropical North Atlantic Ocean and brings warm, humid conditions; the other forms in the Canadian Arctic and brings cold, dry weather. Their tracks are indicated in Figure 1.1. They bring periods of settled, if very different, weather, when conditions change little from day to day. So we can think of them as calms between storms.

AIR MASS TYPES

Any part of the earth's surface where there is a large area of uniform temperature has the potential to serve as a source region for air masses. However, we find that they actually form in the more or less stationary eddies that occur on the edge of the westerlies. An eddy will be created from the air that is left in an area when

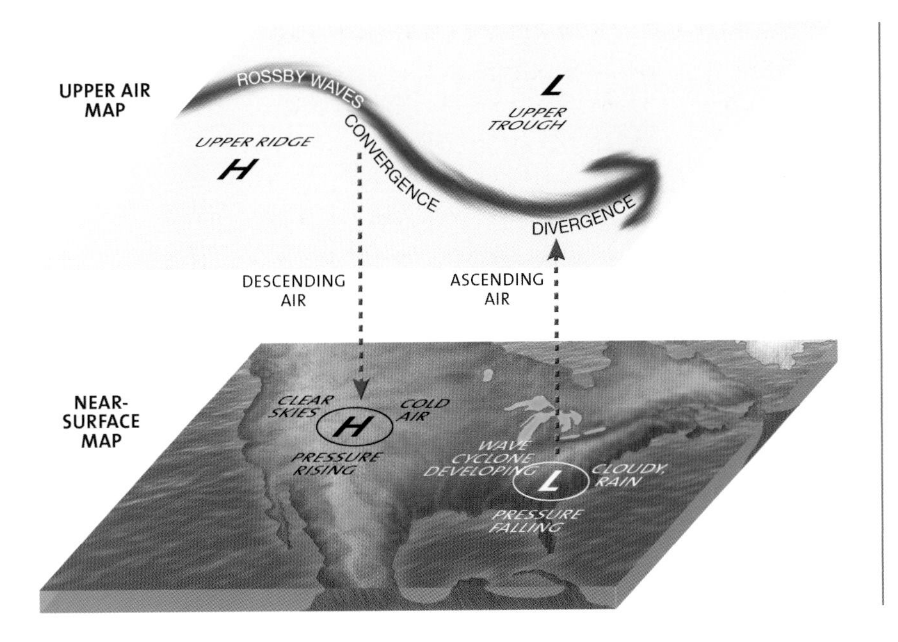

FIGURE 4.7.
Links between the upper air flow and pressure changes—and wave cyclone development—near the surface. Although there is a two-way interaction, the slower and sometimes more predictable changes in the upper atmosphere are commonly used as the main guide to synoptic forecasting.

the Rossby waves change their flow pattern. When this eddy, containing virtually still or very slowly circulating air, sits for several days over the same surface, the area becomes the source region for an air mass. In this source region the air will have its temperature, humidity, and cloudiness character molded by that surface. A cold surface will lead to cold air, and a water surface will produce humid air. The air will slowly become the horizontally uniform "block" we know as an air mass. Eventually changes in the westerly flow may cause that block to move. As it does so, it takes its characteristic weather with it.

Since eddies can form on the poleward and tropical sides of the westerlies and over land or ocean, there are four basic types of air mass (see Table 4.2). The name for each indicates its source region. So does the two-letter symbol (why the first letter is lowercase and the second is uppercase is lost in the mists of time, but at least it makes a readily identifiable symbol). In the 1930s, air mass movement was used as a major weather forecasting tool. Several dozen air mass types were identified, each with closely specified temperature and humidity limits, and their development and movement were carefully monitored and forecast. These days we have many more—and better—ways of forecasting, but we still use the basic concept and the four types in discussing our weather patterns and their daily sequences.

AIR MASS WEATHER

How do we know when we are under the influence of one or another of the various air masses? Table 4.2 gives several clues, but of course there are many permutations and combinations. In general we are likely to be under the influence of an air mass whenever there is an absence of frontal or storm activity in our area. Often winds are light in an air mass, but that is not always the case.

The warm, moist maritime tropical (mT) air mass is common over our state in both summer and winter (warm here is relative to the average temperatures we would expect in each season). The high humidity often leads to a somewhat hazy blue sky. The mT air mass tends to drift slowly into our area and then remain stationary for several days. That drift, however, is responsible for the seasonal contrasts. In summer the land surface is warmer than the sea. The air mass is warmed as it moves inland, vertical motions are encouraged, and cumulus clouds are to be expected. The early afternoon development of these clouds, culminating in thunderstorms in the late afternoon, is a sure sign of a summer mT air mass. In winter the air mass blows off warm water onto relatively cold land. Uplift is slower and more widespread, so stratus clouds often dominate, and any rain from mT air is likely to be cold drizzle.

The source region for the continental polar (cP) air mass is the Canadian Arctic. The earth's surface there, especially in winter, is very cold, and the air mass will be cooled from below. This produces not only cold air near the surface but also the stable atmospheric conditions that impede cloud formation. Further, this cold area has little evaporation, so it also becomes a region of low humidity. When this air mass is swept southeastward to pass over our state, it brings its cold, dry character with it. If there is snow cover between Canada and us, the air never gets a chance to warm during the journey, and it is exceptionally cold when it arrives. Most of our low temperature records have been set in such circumstances. More often, however, in moving south it passes over an increasingly warm, snow-free surface. This warming encourages some convection, and we may see a few small clouds arrive with the cold air. However, low humidity and a deep blue sky are likely. Since this cP air mass often moves into our state rather quickly, wind speeds can be quite high. Although similar dry, cool, and clear conditions are associated with the summer cP air, the position of the polar front ensures that it stays well to the north of us, so we have little experience of it.

The direct influence of maritime polar air is rare in our state, since its source region is the northern North Atlantic and to reach us it must penetrate inland against the westerlies. Most of the time our mT air is incorporated into the flow of a wave

TABLE 4.2. Characteristics of Air Masses Influencing North Carolina

AIR MASS	SYMBOL	TEMPERATURE	HUMIDITY	CLOUDS	FREQUENCY
Continental polar	cP	cold	dry	small cumulus	common in winter
Maritime polar	mP	cool	moist	altostratus	infrequent
Continental tropical	cT	hot	very dry	haze	rare
Maritime tropical	mT	warm	wet	winter: stratus summer:stratocumulus	common dominant

cyclone moving northeastward across the state. It is swept around the north side of the cyclone and arrives from a generally northerly direction as a cool, moist airstream with rather thick stratus clouds. Sometimes mT air is introduced over the coastlands as a strong northeast wind as part of the circulation of a nor'easter, a feature we look at in the last section of this chapter.

We see continental tropical (cT) air in North Carolina rather infrequently. The local source region is the deserts of the southwest. But the mountainous terrain there influences the westerlies so that they rarely force the cT air to move anywhere. When movement does occur, we get very dry, hazy, and hot conditions.

Wave Cyclones: Stormy Weather

Most of our stormy weather is produced by wave cyclones embedded in the westerlies. These regions of ascending air with cyclonic (counterclockwise) flow originate, in association with the jet stream and the polar front, on the Rossby waves. Hence the name. However, sometimes they are also called *frontal cyclones*, because weather fronts are embedded in them, and it is the fronts that are responsible for much of the weather that these storms bring.

FRONTS AND THEIR WEATHER

Fronts are long, narrow regions usually depicted as a line on a weather map with one side cold and the other warm. Although they can vary in length from 1 to 1,000 miles, they tend to have three distinct size ranges. Shortest are local fronts, such as the sea-breeze front we shall consider in Chapter 6. Longest are the polar fronts we have already discussed. In between, often stretched for a few hundred miles, are fronts associated with wave cyclones. These are usually labeled as stationary when they are not moving, warm when they move such that the weather gets warmer as they pass overhead, and cold if the weather gets colder. The final type, occluded, occurs in a wave cyclone when warm and cold fronts interact. All have the same basic character. They are created when warm air and cold air flow toward each other. The two airstreams rarely meet head-on; most of the time they converge like traffic merging on a freeway. There is some mixing of the converging air, so the boundary between the two, the front itself, is a zone rather than a precise line. But most of the air does not mix. Rather, the warmer, less dense air rides up over the colder, denser air. So the front also slopes upward into the atmosphere, "tilting" toward the cold side. Most important for the weather, clouds and often rain are associated with this vertical motion and the sloping front. The cold front shown in Figure 4.2 has cold air from the west forcing warm air over the ocean to rise, creating clouds somewhat ahead of the front.

A WAVE CYCLONE OVER NORTH CAROLINA

Wave cyclones and their weather commonly develop in a fairly predictable way. For us in North Carolina, most are started when the Rossby waves develop a southern bend to our west, often around the lower Mississippi valley. The sucking of air away aloft creates a low pressure area at the surface (see Fig. 4.7). This creates low pressure and cyclonic motions at the surface (see Fig. 4.8a). When the polar front is in the region, this motion creates a bend in the front at the center of low pressure. This is the beginning of the wave cyclone. This situation may persist for several days; then changes in the upper airflow may simply cause the uplift to decrease or cease, and the low pressure center will vanish.

Much more interesting, of course, is the other possibility: the upper wave pattern persists and the cyclone begins to develop. Then the low pressure center deepens and moves along within the general flow of the westerlies. The two parts of the polar front, which now seem to hang south from the low pressure center, become distinct warm and cold fronts (see Fig. 4.8b). Characteristic weather patterns begin to occur (see Fig. 4.9). To the north of the center there is a region of general uplift, and the air spirals inward and converges. A broad area of stratus clouds, possibly with rain or drizzle, is most likely. If we are to the south of the center as the cyclone passes overhead, however, there is often a very distinct sequence of weather

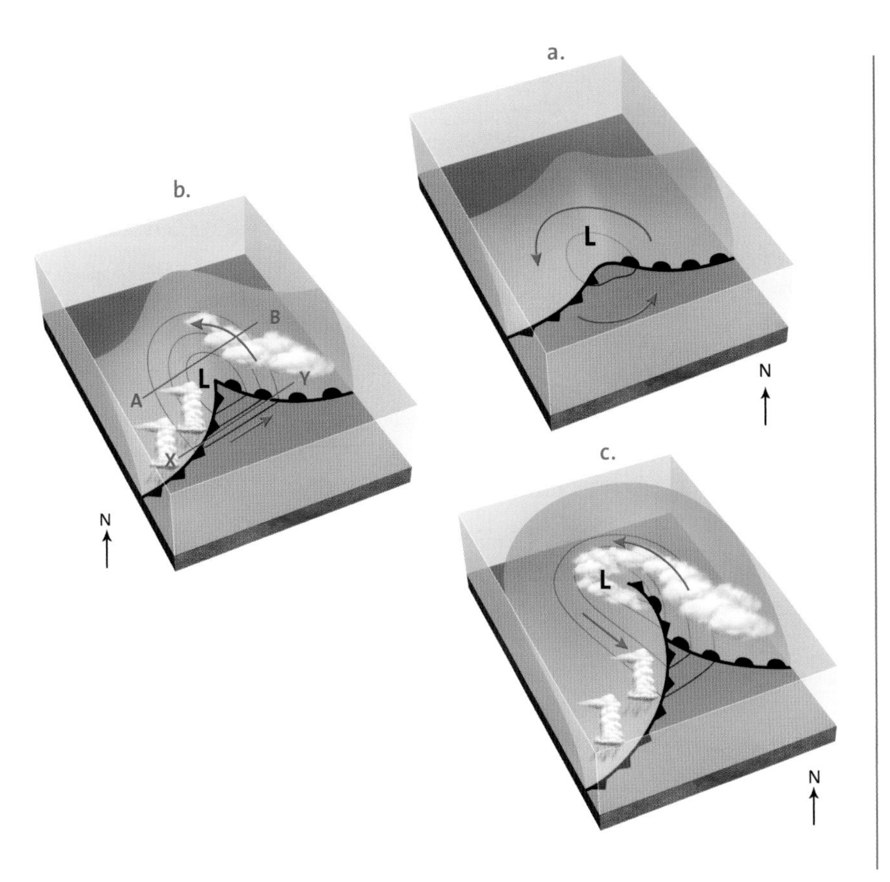

FIGURE 4.8. *Life cycle of a typical wave cyclone. Each stage occurs one or two days after the preceding one. At each stage, the cyclone center will have moved within the planetary waves and may be 500 miles to the east or northeast of the preceding position. Not all cyclones will go through the whole cycle, while the speed of development and movement will vary within a single storm.*

events. In this situation the first sign of an impending wave cyclone is cirrus clouds approaching from the west or southwest. As these high clouds move above us, we see progressively lower clouds coming in from the west. By the time stratus clouds arrive overhead, some rain is likely. As the rain clears and the clouds pass to the east, the relatively cool southerly airstream is replaced by warmer, more westerly air. The warm front has approached and passed us. We are now in for a few hours of warm, humid conditions. This warm sector is probably the remnants of an mT air mass, so some clouds and even isolated showers are possible. The final stage of cyclone passage is heralded by the approach, again from the west or southwest, of a line of cumulus or cumulonimbus clouds. The cold front is approaching. Temperatures drop rapidly; the wind shifts to a more northerly direction. A band of rain, often a short-lived, intense downpour, occurs. The clouds move away to the east, the sun comes out, and the wave cyclone has passed (see Fig. 4.10). Although this appears to be a very precise sequence, it happens often enough that when you

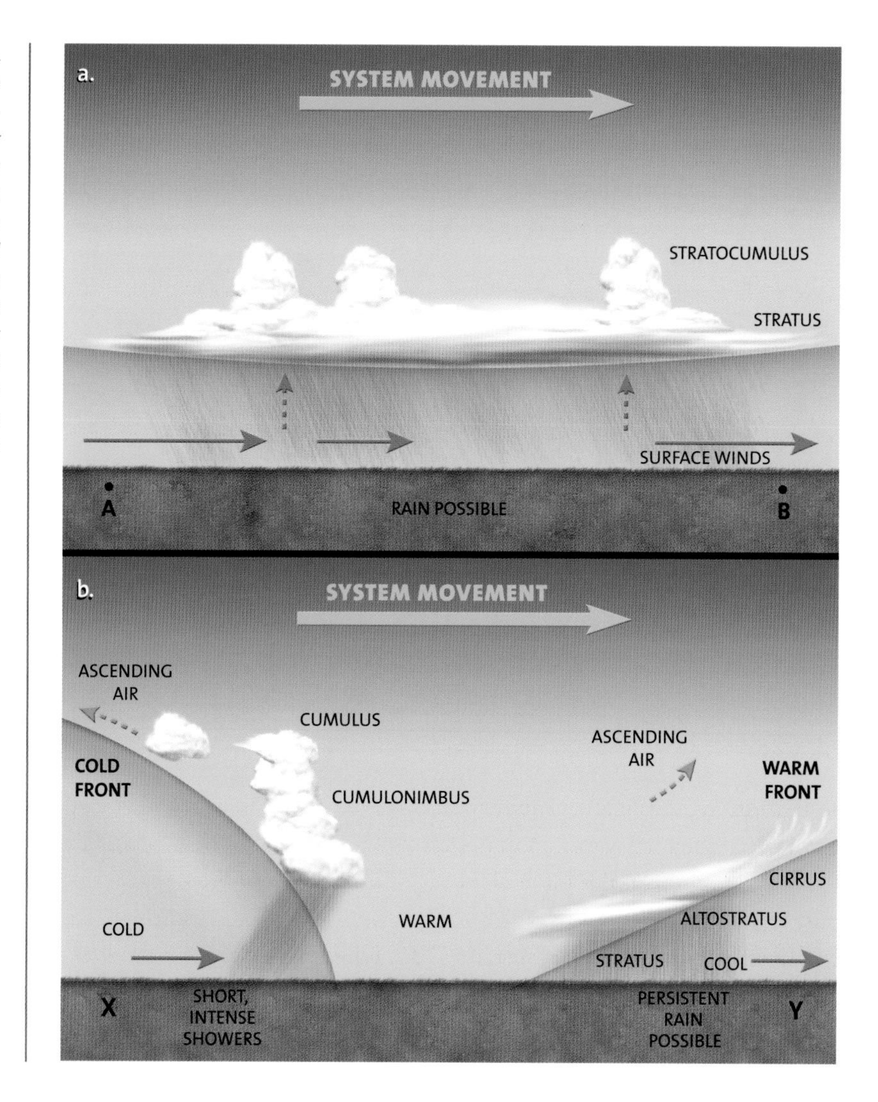

FIGURE 4.9. *Cross section showing the common weather associated with a wave cyclone in North Carolina in (a) winter and (b) summer. The locations of the cross sections for 4.9a (AB) and 4.9b (XY) are shown in Figure 4.8b.*

see a wave cyclone approaching with its center to our north, you can use it as a forecast. You should be right often enough to impress colleagues or family.

In the final stage in the wave cyclone's life (see Fig. 4.8c), the cold front catches up with the warm front and an occlusion, where one front lifts the other completely off the ground, takes place. This is a time of general uplift around the system, so there is widespread cloudiness and, often, prolonged heavy rain. By now, however, the cyclone is likely to be moving under the northern bend of a Rossby wave, an area where the vertical motions are downward. So the low pressure center gets filled in, and the cyclone vanishes.

FIGURE 4.10. *The line of cumulus clouds moves away as a cold front leaves Craven County for the Atlantic Ocean. Intense rain is giving way to cold, sunny weather.*

OUR VARIABLE WAVE CYCLONE WEATHER

There is a strong seasonal and annual variability in the influence of wave cyclones on North Carolina. However, most are formed in the region roughly bounded by the Rocky Mountains, the Mississippi River, the Canadian border, and the gulf coast. The exact position of their formation, development, and track will depend on the various aspects of the westerlies. But we most often see them after they have traveled at least a thousand miles from their breeding ground, when they are approaching the mature phase of development and resemble the patterns of Figure 4.8b or 4.8c and Figure 4.9.

In winter the low pressure center of the cyclonic circulation tends to cross directly over our state from southwest to northeast (see Fig. 4.9a). The general ascending motion of the air gives a region of widespread clouds. Sometimes these are mainly altostratus, giving light rain or drizzle; more often they are stratus, and we get heavier rain. In either case, precipitation is likely to continue for several hours and sweep over virtually the whole state. Most of our winter precipitation comes from these systems. The westerlies tend to be strong, the system is fast moving, and wind speeds are high (Box 4.1). Most of the time the wind is from the southwest. Temperatures are likely to be lower after the cyclone passes.

In summer, conditions are very different (see Fig. 4.9b). The storm center tends

BOX 4.1. MARCH 1993: THE "STORM OF THE CENTURY"

Not all destructive and deadly storms are hurricanes or tornadoes. The "storm of the century" was a frontal system of the type we see passing over us several times a month. It began, innocently enough, on March 10 as an unremarkable low pressure system in the area of the southern plains and the western Gulf of Mexico. But it intensified very rapidly as it started to move northeastward on March 12. It moved fast, directly toward us. By the time it passed over the state on March 13, the pressure pattern was more like that of a hurricane than a frontal system. The center set new low pressure records at Asheville, Charlotte, and Raleigh-Durham, and wind gusts of more than 50 miles per hour were observed for much of the state.

In the west, blizzard conditions prevailed. Winds reached 60 miles per hour, and two feet of snow fell in the valleys. Many peaks had almost double those snow amounts, while winds approached 100 mph. Virtually all activity was brought to a halt, thousands of people were stranded without power, and there was considerable damage, especially to roofs, which either blew away or collapsed under the weight of the snow. In the east, wind damage was also widespread, and falling trees caused two fatalities and several injuries. The high winds also drove ocean waters ashore, giving storm surge flooding all along the coast.

As the system moved away across Virginia, cold air swept southward around the departing low pressure center. The high winds and low temperatures combined to create wind chills equivalent to temperatures below -20°F throughout the mountain region. Seven deaths were attributed directly to exposure, while at least seven others appeared to be weather-related. These included snow-shoveling heart attack fatalities. Not counted were several fatalities associated with the 2,700 traffic accidents during the storm. Overall, the storm left at least 300,000 homes without power and 160,000 people snowbound.

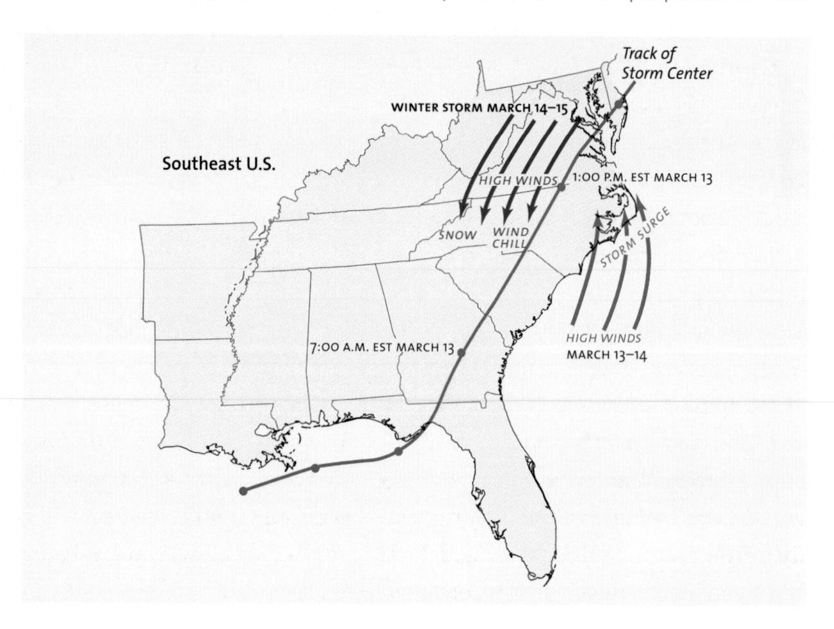

FIGURE 4.B1. *The track and impacts of the storm of the century*

to be some distance to our north, and we frequently get the distinct sequence of weather events described in the previous section. Furthermore, the westerlies are less vigorous than in winter. The systems are slower moving and tend to give less total rainfall, although the thundershowers associated with the cold front can be very intense. Nevertheless, for much of the state away from the mountains, wave cyclones commonly bring only about one-half to three-quarters of our total summer precipitation. The rest comes from the much more randomly distributed thunderstorms within the mT air masses.

In some summers, the cyclone tracks are greatly influenced by the Bermuda high, the high pressure region that usually sits east of us, centered on the island that gives it its name. But in some summers the high pressure region expands or simply moves westward (see Fig. 4.11). Then it sits over our Coastal Plain. This high pressure acts as a barrier to wave cyclone movement, forcing the storms to pass much farther north than normal. We get no rain from them. Further, the high pressure region also discourages the development of thunderstorms from the mT air masses, so we loose that water source as well. While the Bermuda high is likely to take up this westerly position for a few weeks in any summer, in some years it persists for a month or more. Drought then becomes a possibility. Indeed, the worst droughts in the agricultural regions of the Coastal Plain have occurred when the situation persisted for most of the summer. In those cases the droughts have been ended either by the passage of a hurricane or by the onset of fall, both of which rearranged the whole atmospheric circulation of the area.

In winter, one effect of changes in circulation involves the position of the southern bend in the planetary waves and the development of winter storms. Sometimes there is a strong cold air outbreak in the midsection of the nation, and cold air sweeps across the plains, possibly reaching almost to the gulf coast. This forces any developing wave cyclone out over the warm waters of the Gulf of Mexico. Evaporation puts a lot of moisture into the air, and this provides energy to invigorate the cyclone. Further, the temperature contrast between the very cold land and the very warm water produces strong fronts. As a result, a strong wave cyclone guided by the jet stream moves northeastward out of the gulf and across the southeastern United States. The vigorous circulation creates lots of clouds and precipitation, and the very cold air ensures that the precipitation can fall as snow. Then there may be snow in the Florida panhandle, and there certainly will be some through Georgia and the Carolinas. Often the system has lost much of its vigor and warmed up somewhat by the time it gets into Virginia, where it will give only a little rain and no snow. Although this pattern seems backward, it happens often enough to be readily understood and forecast.

FIGURE 4.11. *Atmospheric conditions creating summer drought*

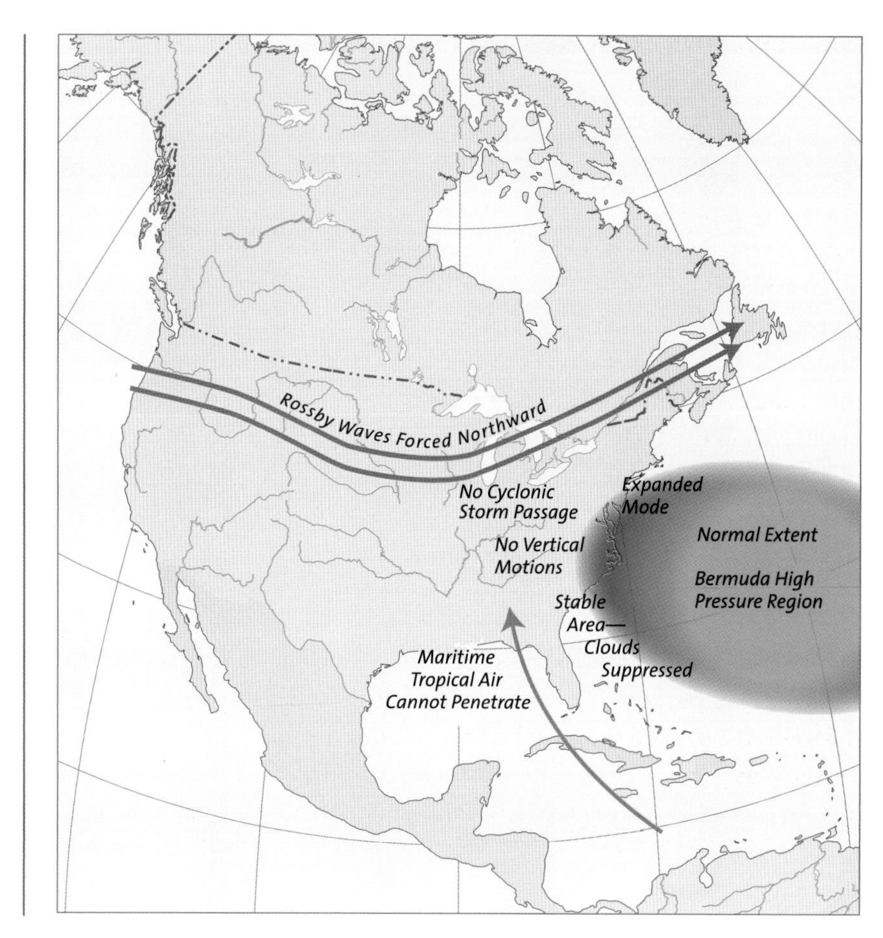

The development and movement of wave cyclones is not always as simple as we have implied here. In particular, the Appalachian Mountains can provide a barrier to movement. Experience has shown that when a storm approaches our state from the west and encounters the mountains, it can (1) change track and move northeast along the west side of the mountains, more or less following the Ohio Valley; (2) be slowed, or possibly stopped for a while, by the mountains, but continue on a virtually unchanged track; (3) follow a track along the crest, or somewhat east of the crest, of the mountains; or (4) ignore the mountains and continue as if they did not exist.

Which scenario will occur depends on many things, some of which we are not sure we understand. Certainly in order to forecast the tracks, in the mountains or anywhere else, we need to know the links between the jet stream, the planetary waves, the wave cyclones, and the air masses.

WAVE CYCLONES CREATED IN NORTH CAROLINA: NOR'EASTERS

In addition to the wave cyclones that arrive over our state from the west, there are similar storms that originate over our coastal areas. These are often called Atlantic lows or Hatteras lows. They are particularly important in winter. At that time there is a strong temperature contrast between cold land and warm water, so a vigorous front can develop. It seems that the creation of this front is also encouraged by the configuration of the shoreline around Cape Hatteras, while it is suspected that the contrast in ocean temperatures north and south of the cape—partly associated with the confluence of the Gulf Stream and the Labrador Current—and the mixture of ocean, sound, and barrier island may also have roles to play. In any event, around Hatteras it is possible for very deep (extreme low pressure) systems to develop very quickly. Most of the time the fronts within the system are much less significant than the winds. The steep pressure gradients give strong turbulent conditions throughout the area.

Sometimes this low remains in the Hatteras area for a considerable time, leading to a true nor'easter. The cyclonic circulation around the low brings winds from the northeast into North Carolina's northern coastlands. The winds push the water in the same direction. The high winds and little system movement give time for the waves to build to a considerable height. Along the coast, conditions approaching those usually associated with hurricanes can occur. Beach erosion is common, and high winds can cause property damage. In the sounds, shoreline erosion is likely on the southern sides, as the wind pushes the water southward and creates unusually high water. In the north, low water occurs, often causing the beaching of unprepared watercraft.

At other times the Hatteras low not only develops quickly but also moves away rapidly. In winter it is common for the following sequence of events to occur: A small and often intense low pressure system develops in the lee of the Rocky Mountains. This "clipper" storm moves rapidly across the upper Midwest, bringing high winds and blowing snow. It tracks across the Great Lakes to New England, well to the north of North Carolina. Immediately afterward, perhaps stimulated by the passage of that storm, an Atlantic low develops exceedingly rapidly and moves northward to the New Jersey coast, then turns to follow the previous storm. The second storm is often vigorous enough to create blizzard conditions in New England.

chapter 5

Severe and Hazardous Weather

Most of us find the atmosphere to be most interesting, exciting, and perhaps terrifying when severe weather occurs. Many professionals originally became interested in meteorology because they wanted to know more about severe weather, particularly hurricanes and tornadoes. But hazardous weather also includes winter storms, floods, droughts, dangerously hot—and sometimes cold—spells, and increasingly, weather patterns that allow air pollution to become a health hazard. Considering the variety of conditions we experience, we are not short of interesting weather in North Carolina.

Hurricanes

On any day in North Carolina between June and December there is in the background, and maybe the foreground, the threat of a hurricane. Hurricanes have been hitting us, usually along the coast but sometimes far inland, since long before human records were kept. Over the years our understanding of these storms and our ability to predict them have increased. While we may not yet have lessened the wind and water damage a severe hurricane can bring to our state, we have reduced the number of lives lost because people were unprepared for them.

HURRICANE FRAN

Every hurricane has its own distinct personality, so it is not really possible to consider an “average” hurricane. But Fran of August–September 1996 came closer than most to being typical, and we can use it to look at how hurricanes form, develop, move, and eventually die (see Fig. 5.1).

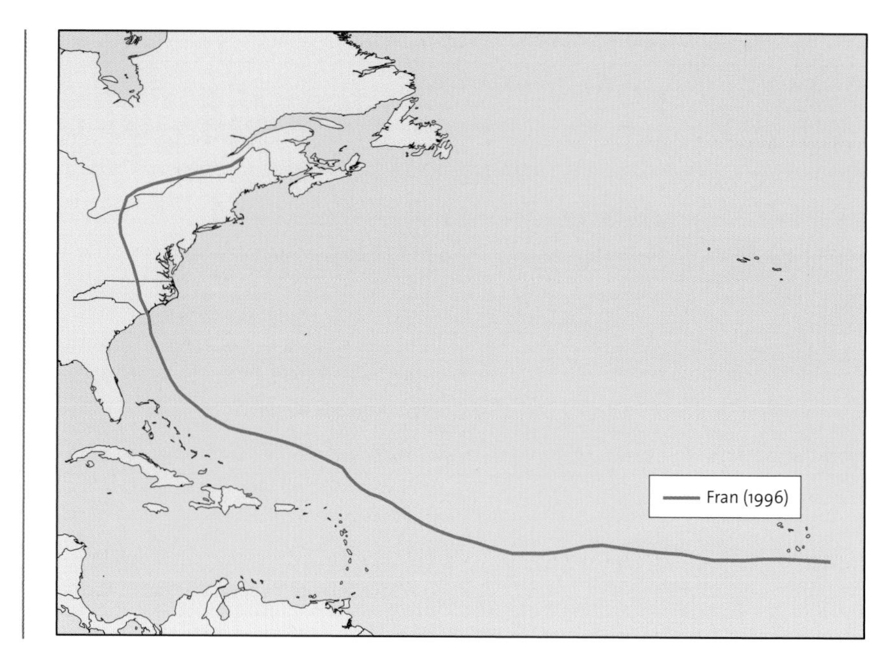

FIGURE 5.1. *The track of hurricane Fran, August–September 1996*

THE ORIGIN OF THE STORM

The majority of our hurricanes appear to start as small disturbances in the airflow off the African coast. The origin of Fran was a tropical wave that moved westward off the African coast on August 22. Convection associated with this wave gave a continuously rising plume of air that started near the surface, went all the way to the tropopause, and spread out aloft. The rising plume began to spiral in a cyclonic (counterclockwise) motion throughout its whole depth, the central pressure decreased, and what was originally mainly a series of thunderclouds became a tropical depression (see Table 5.1). This occurred on August 23 when the system that became Fran was just south of the Cape Verde Islands.

The westward movement of atmospheric waves off the African coast and the development of thunderstorms occur fairly often, but few of these disturbances become hurricanes. There must be special factors at work to transform a seemingly minor feature into a major event. We do not know all of the factors involved. We know that a deep pool of hot ocean water is required to provide the energy for the system. We know that the wind speed and direction cannot change much with height or the budding hurricane will be sheared apart. We know that a hurricane cannot form too close to the equator because the effect of the spin of the earth is

TABLE 5.1. National Weather Service Classification of Storm Systems

FEATURE	WIND SPEEDS	COMMENTS
Tropical depression	less than 39 mph	Circular flow around a low pressure center with the potential to become a hurricane, or the remnants of a decaying hurricane.
Tropical storm	39–73 mph	Well-organized circular system with a marked low pressure center having high potential for intensification into a hurricane, or a hurricane-like storm occurring when a hurricane begins to decay and wind speeds decrease.
Subtropical storm	39 mph or greater	Origin 25–30°N (or S) with strong circular flow, but will not develop into a hurricane.
Hurricane	74 mph or greater	Well-organized circular storm divided into various intensity categories (see Table 5.2).
Extratropical storm	no wind speed criteria	Has a circular form similar to a hurricane but less intense and outside the tropics. May include weather fronts and can become intense. The wave cyclones and nor'easters of North Carolina's routine weather are examples.

not strong enough there to get the air going into the required circular motion. But when conditions are right, a disturbance off the African coast becomes organized as a Cape Verde storm and may, like Fran, develop into a major hurricane.

MOVEMENT AND DEVELOPMENT OVER THE OCEAN

Tropical depression Fran drifted westward at about 15 knots for a few days, changing little in intensity. Then on August 27, when it was 900 nautical miles east of the Lesser Antilles (and still more than 2,000 miles from the North Carolina coast), for reasons not well understood, it intensified to a tropical storm and began to head northwest. Two days later it became a hurricane. But as it passed just north of the Leeward Islands, Fran's wind speed dipped slightly. It returned to tropical storm status for a while and began to move rather slowly. Soon it was back to hurricane strength, moving at around 10 knots northwestward toward the U.S. coast.

Fran's general track had been fairly common for a Cape Verde storm. The exact route was probably influenced strongly by the preceding hurricane, Edouard, which had taken a similar track only a few days earlier. However, as Edouard moved away, it was not clear whether Fran would continue to be influenced by the old hurricane or by other features in the midlatitudes. That made the forecast of the subsequent track very uncertain. Hurricanes, when they are near where Fran was on August 31, often change direction from heading west to a more northerly track and even to a northeast one before they die out. But how fast and exactly where they make this shift often depends on midlatitude conditions. In Fran's case there was a ridge of high pressure behind Edouard that encouraged Fran to continue moving westward rather than follow Edouard north to the west of Bermuda. But there was also a region of low pressure developing in Tennessee that had the potential to steer the system toward the north.

THE MATURE STORM HITS NORTH CAROLINA

By September 1, Fran was under the influence of the high pressure ridge that separated it from the decaying Edouard. Fran tracked to the northwest at about 10 knots. Soon the influence of the low pressure region over Tennessee began to have an effect. Both the high and the low were encouraging airflow from the south, and this helped to guide the storm into a more northerly track between them. The flow also encouraged the development and intensification of the system. By September 4, Fran had become a category 3 storm (see Table 5.2). It reached a maximum intensity on September 5, with surface pressure of 946 millibars and sustained winds of 105 knots when it was 250 nautical miles east of Florida.

At this stage Fran was a mature hurricane with a well-established structure. As is typical for mature hurricanes, there was a marked circular airflow around the low pressure center at the eye (Fig. 5.2), while separate regions of ascending and descending air led to distinct cloud and rain features.

As Fran entered and passed over the state, it had the characteristic weather sequence associated with a mature hurricane. The approach is heralded by the arrival of the thin veil of cirrus clouds, often with very little wind. Slowly these rather pleasant conditions are replaced by increasingly thick clouds, stronger winds, and rain. The increase is not uniform because the inward-spiraling airflow creates cloud bands (see Fig. 5.2). As a band passes overhead, the wind and rain increase and then taper off, only to be replaced by more intense conditions as the next band arrives. There may be several bands, although a surface observer may not distinguish them in the general windy, rainy conditions. As the eye of the storm approaches, the bands get closer together and act more like a continuous

TABLE 5.2. Simpson-Saffir Scale for Hurricanes

SCALE NUMBER (category)	CENTRAL PRESSURE (mb)	WINDS (mph)	STORM SURGE (ft)	DAMAGE
1	≥980	74–95	4–5	Damage mainly to trees, shrubbery, and unanchored mobile homes.
2	965–79	96–110	6–8	Some trees blown down; major damage to exposed mobile homes; some damage to roofs of buildings.
3	945–64	111–30	9–12	Foliage removed from trees; large trees blown down; mobile homes destroyed; some structural damage to small buildings.
4	920–44	131–55	13–18	All signs blown down; extensive damage to roofs, windows, and doors; complete destruction of mobile homes; flooding as far as 6 miles inland; major damage to lower floors of structures near shore.
5	<920	>155	>18	Severe damage to windows and doors; extensive damage to roofs of homes and industrial buildings; small buildings overturned and blown away; major damage to lower floors of all structures less than 15 feet above sea level within 1,000 feet of shore.

feature. Tornadoes may occur. The wind, rain, and clouds reach their culmination as the eyewall arrives. This thick, solid cloud mass is replaced by the eye itself, which brings a virtually calm period when it is possible to look up and see a blue sky or a starry night. The eye is small, only about ten miles across. Usually, the more intense the hurricane, the smaller the eye. So the passage of the eye gives only a short respite. The sequence of wind and rain bands is then repeated, with the wind blowing from the opposite direction.

Although Fran had weakened somewhat in moving north from the waters off Florida, it still arrived in North Carolina as a category 3 storm with a central pressure near 954 millibars and sustained surface winds of 100 knots. It was moving forward at about 15 knots when it hit the lower Cape Fear region late

FIGURE 5.2. *A mature hurricane seen in (a) cross section and (b) plan views*

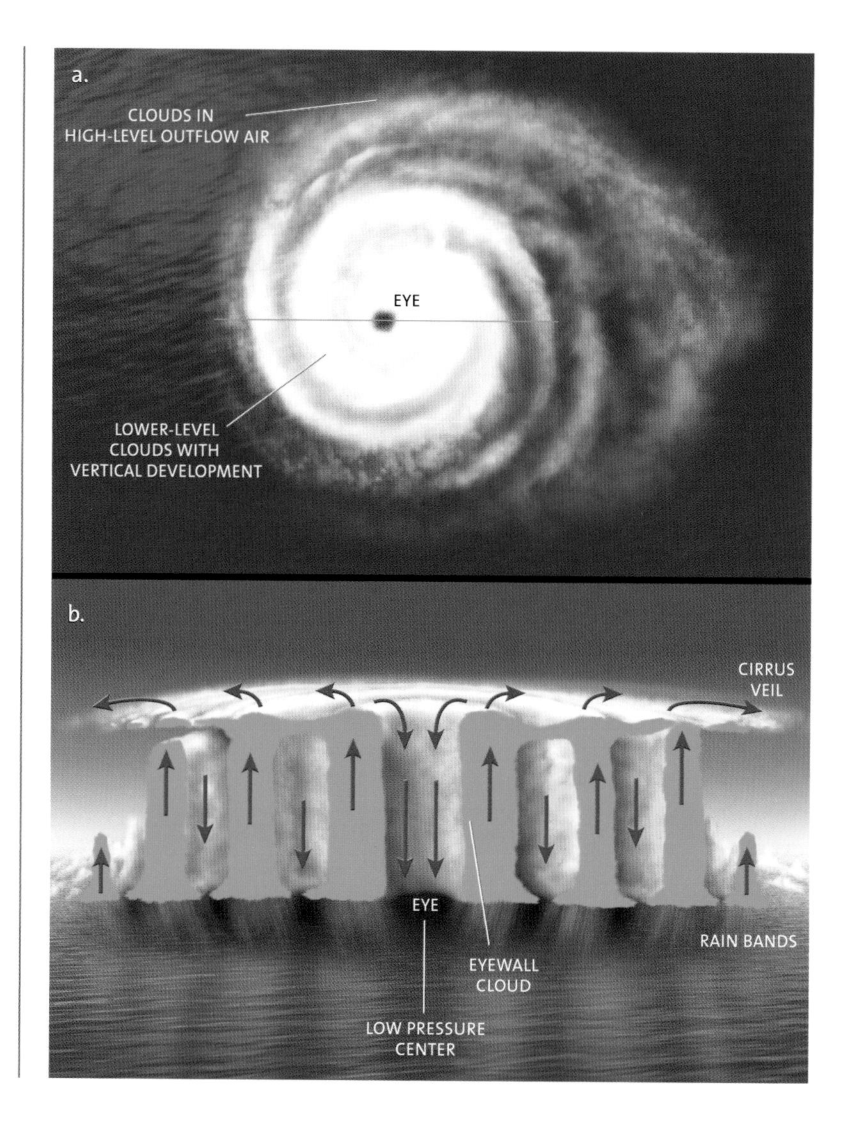

at night on September 5. Winds of hurricane force hammered the coast from Brunswick to Carteret counties.

THE STORM IN NORTH CAROLINA

Fran moved quickly, and the eye was over the state for only about twenty-four hours. In many respects it was similar to hurricane Hazel in 1954, the last major storm to pass directly across the eastern Piedmont and the only category 4 storm to make landfall in our state during the twentieth century (see Box 5.1). Fran brought relatively little rainfall; most of the region got around 6", although some coastal areas had 12". There was some flooding on the major rivers, but most damage was associated with wind, and particularly with falling trees. The ground was soaked from the effects of a wet summer, so trees were easily uprooted by the wind. Most of the areas affected had not had such high winds since Hazel, and a tremendous amount of development had taken place. Many houses were situated among mature trees that were highly vulnerable to wind effects. Property damage exceeded $2 billion in the northern Piedmont alone. Further, most deaths were associated with falling trees and collapsing chimneys and other structures, but several deaths resulted when motorists attempted to drive through flooded streams.

MOVING AWAY TO THE NORTH AND DYING OUT

Traditionally, hurricanes rapidly weaken once they are over land and are cut off from the evaporating water that is a major energy source for them; they are also slowed by the extra friction created by the rougher surface of the land. Even over water, they usually decay as they move north over colder oceans. Fran, in fact, weakened rapidly and was a tropical storm by the time it reached central North Carolina. In Virginia it was a tropical depression. As the storm weakened, it produced even more rainfall, and floods in Virginia and Pennsylvania were extensive. The system was, however, strong enough to continue across Lake Ontario, and it merged with a frontal system in southern Ontario on September 10.

NORTH CAROLINA HURRICANES AND THEIR IMPACTS

More than fifty hurricanes had an impact on North Carolina during the twentieth century (see Appendix D). The exact number is hard to determine, since our observational systems improved tremendously during the century. We can now, using satellite observations and aircraft reconnaissance, for example, detect storms far offshore that are causing coastal erosion, or we can pinpoint disturbances well to the west and south that are creating floods in North Carolina. One hundred years ago we probably would not have recorded these events as hurricane related. Nevertheless, we can estimate the frequency of hurricanes by decade, and it is clear that some decades had lots of hurricanes while others were almost hurricane free (see Table 5.3). The alternation between quiet and active periods suggests that there might be a cyclic tendency. If this is so, the indications are that the early part of the twenty-first century is likely to be active for hurricanes in the Atlantic basin.

BOX 5.1. OUR CATEGORY 4 HURRICANE: HAZEL IN 1954

For most people growing up or arriving in North Carolina in recent decades, hurricane Hazel in 1954 is probably the most talked-about weather event, even though the hurricane was only in the state for a few hours. Although it may now be overshadowed in recent memory by Fran or Floyd, hurricane Hazel was still the only category 4 storm to hit us for at least 100 years.

Hazel was one of the fastest-moving storms in history. For most of the state, the morning of October 15, 1954, started simply as a continuation of a dry and unseasonably hot spell. High temperatures in Raleigh the day before had reached 98°F, and the overnight low had been in the 70s. To the south was hurricane Hazel, rapidly heading north. By 7:30 A.M. it was some 200 miles off our coast, and the beaches near the North Carolina/South Carolina border were getting high surf and seeing approaching clouds. Conditions quickly deteriorated, and in the middle of the morning the eye passed over Calabash. Soon after noon it was just west of Raleigh, and before nightfall it was through our state, past Virginia and Pennsylvania, and was located in upstate New York. Behind the storm came a period of weather that was much colder than normal.

Damage and casualties in inland regions were mainly a result of the extremely high winds; there were several reliable reports of winds in excess of 150 mph. These winds were associated both with the flow around an eye with an exceptionally low pressure and with the very rapid forward speed of the whole storm. Although rain was often intense, the rapid forward movement meant that not much precipitation fell on any one area, so rainfall totals were relatively low. Much of the rain that did fall was absorbed by dry soils, so there was little inland flooding. On the coast the storm surge reached record heights, with water eighteen feet above normal sea level at Calabash. Seawater penetrated a good distance inland, and property damage and shoreline erosion were extensive.

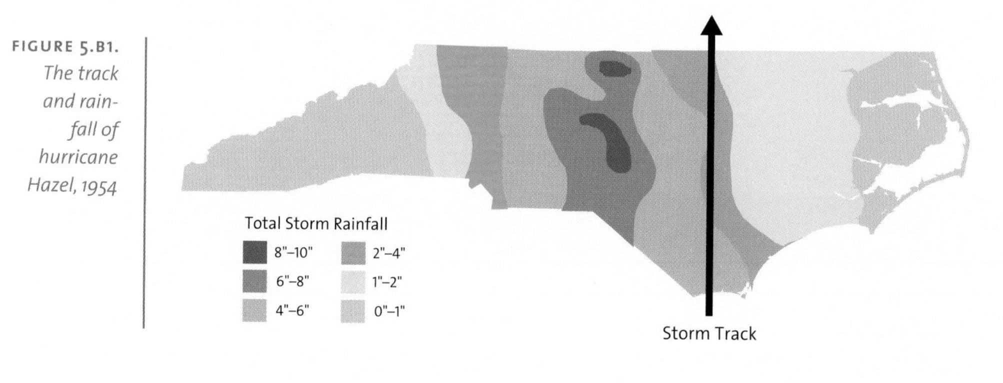

FIGURE 5.B1. *The track and rainfall of hurricane Hazel, 1954*

The official hurricane season of the National Weather Service (NWS) starts on June 1 each year. The earliest arrival for us was a storm on June 3, 1826. The earliest in the twentieth century was Alma, which was offshore of Cape Hatteras on June 11, 1966. Tropical depressions can arrive even earlier; one, also named Alma, was

TABLE 5.3. Frequency and Intensity of Hurricanes Influencing North Carolina, by Decade

CATEGORY	1901	1911	1921	1931	1941	1951	1961	1971	1981	1991
4	-	-	-	-	-	1	-	-	-	-
3	2	-	-	1	1	4	-	-	2	2
2	-	-	-	2	-	2	-	-	1	3
1	6	3	5	3	5	2	4	1	1	2
Tropical storm*	-	3	-	1	3	3	3	6	4	5

*The numbers for tropical storms are estimated in the earlier years.

off Cape Hatteras on May 26, 1970. However, the ocean waters in May are still fairly cool, and depressions have insufficient energy to turn into hurricanes, so Alma in 1970 never progressed beyond the depression stage. Indeed, hurricanes in June and July are rare for us. August and September are the most active months (see Fig. 5.3). We cannot relax until mid-December, however, since the latest hurricane to come ashore was a category 1 storm that passed over Emerald Isle and moved northeast toward Norfolk on December 1, 1925.

Every hurricane is unique and seems to have a distinct personality, so it is no wonder that we give them human names (see Table 5.4). A wonderful review of the individual hurricanes influencing North Carolina, emphasizing mainly those that affected the coast, has been written by Jay Barnes. Here we have a few examples to demonstrate the meteorological aspects and impacts. We only hint at the human dramas involved with almost every hurricane.

Hurricanes were not named until 1953. Prior to that they were known by their date of arrival. Thus there was a major "Labor Day storm" that devastated the Florida Keys in 1935. As our observational ability increased, so did the confusion, as two or more hurricanes could occur on the same date. So in 1953 the National Hurricane Center developed alphabetical lists of female names for each year's hurricanes. Names could be reused unless they were associated with particularly severe storms. Following the storm that crossed North Carolina in thirty-six memorable hours in 1954, the name Hazel was retired, and we shall never see another hurricane Hazel. Starting in 1979 male names were added, alternating with female ones. The World Meteorological Organization now maintains six lists that are used in rotation (Table 5.4).

Although each hurricane is unique, there seem to be three major tracks for those that affect our state. One track passes along or parallel to our coastline. Another enters the state along the southern coast and tracks generally northward.

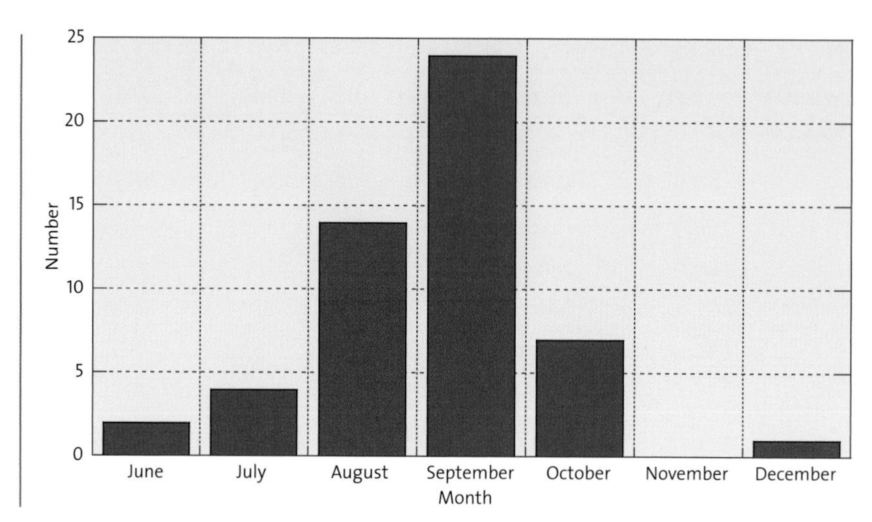

FIGURE 5.3. *Number of hurricanes that affected North Carolina each month during the twentieth century*

On a less common path, storms from the Gulf of Mexico move northeastward and enter the state from the southwest (see Fig. 5.4).

COASTAL TRACK

In sheer numbers, hurricanes taking the coastal track dominate. The general track is an arc from the south toward the northeast (see Fig. 5.4). An actual track may be entirely over the Coastal Plain, entirely offshore, or varying between the two locations. They all cause wind and rain on the Coastal Plain, but few are likely to provide enough rain in the headwaters of the state's eastern rivers to cause much flooding. But they all create a storm surge. The high winds north of the approaching storm blow shoreward over the ocean and literally push the ocean waters onshore. Additionally, the low pressure at the storm's center causes the height of the ocean surface to "dome" upward slightly, so that the waves of the storm surge are slightly higher than we would normally expect. The result is usually erosion of the coastline, be that composed of a natural dune or a human residence. The surge is often high and strong enough to penetrate inland over parts of the Coastal Plain. As a result, the most common impact of hurricanes in our state is coastal erosion and flooding along the coast.

In this context, two nineteenth-century hurricanes are noteworthy. In September 1845 an offshore storm approached us from the south at a rather leisurely pace. Although we cannot reconstruct the situation exactly, it appears that persistent high winds from the northeast pushed much water into the sounds. When the eye of the storm passed northwestward over Hatteras, the wind direction shifted, and

TABLE 5.4. Names for Atlantic Hurricanes

2002	2003	2004	2005	2006	2007
Arthur	Ana	Alex	Arlene	Alberto	Andrea
Bertha	Bill	Bonnie	Bret	Beryl	Barry
Cristobal	Claudette	Charley	Cindy	Chris	Chantal
Dolly	Danny	Danielle	Dennis	Debby	Dean
Edouard	Erika	Earl	Emily	Ernesto	Erin
Fay	Fabian	Frances	Franklin	Florence	Felix
Gustav	Grace	Gaston	Gert	Gordon	Gabrielle
Hanna	Henri	Hermine	Harvey	Helene	Humberto
Isidore	Isabel	Ivan	Irene	Isaac	Ingrid
Josephine	Juan	Jeanne	Jose	Joyce	Jerry
Kyle	Kate	Karl	Katrina	Kirk	Karen
Lili	Larry	Lisa	Lee	Leslie	Lorennzo
Marco	Mindy	Matthew	Maria	Michael	Melissa
Nana	Nicholas	Nicole	Nate	Nadine	Noel
Omar	Odette	Otto	Ophelia	Oscar	Olga
Paloma	Peter	Paula	Philippe	Patty	Pablo
Rene	Rose	Richard	Rita	Rafael	Rebekah
Sally	Sam	Shary	Stan	Sandy	Sebastien
Teddy	Teresa	Tomas	Tammy	Tony	Tanya
Vicky	Victor	Virginie	Vince	Valerie	Van
Wilfred	Wanda	Walter	Wilma	William	Wendy

Note: Starting in 2008 the lists will be repeated, subject to substitution of retired names for exceptionally severe storms.

westerly or even southwesterly winds arrived. These encouraged a surge of water from the sounds back into the ocean, eroding the dunes of the Outer Banks. On the night of September 7 a new inlet, Hatteras Inlet, was opened. The following day another new one, Oregon Inlet, was formed.

Over half a century later, on August 16–18, 1899, another slow-moving storm arrived from the south. Just offshore it slowed even more and gained strength while passing over the warm coastal waters. It then turned inland and almost certainly became a category 4 storm. The NWS observer at Hatteras recorded winds of 93 miles per hour (mph), with gusts in the 120 to 140 mph range, before the anemometer blew away. Hatteras Island was covered with water. Farther south,

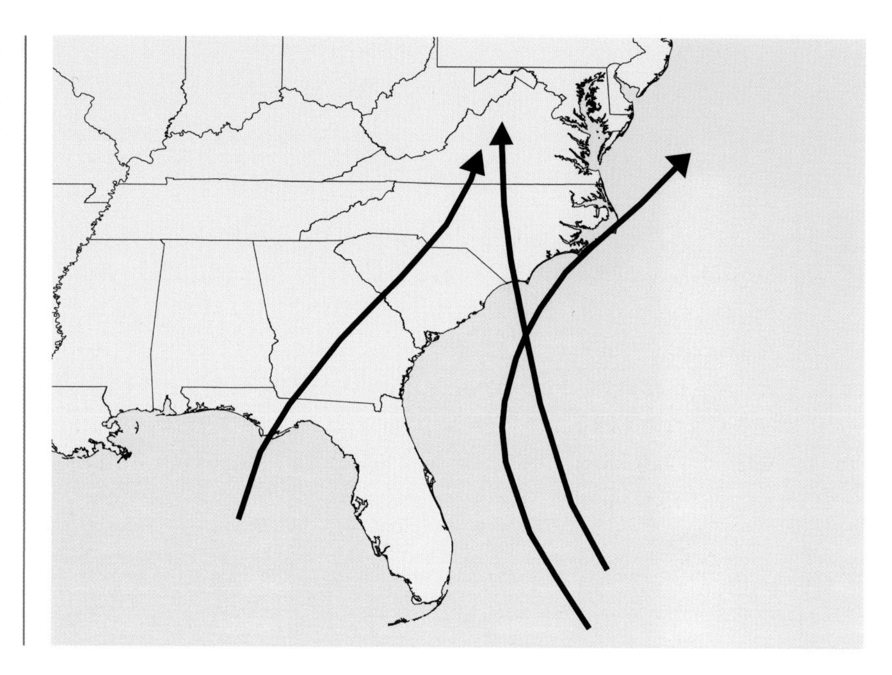

FIGURE 5.4. *Generalization of the three major tracks for hurricanes affecting North Carolina*

settlements on Shackleford Banks were also completely inundated; every building was damaged or destroyed, all garden produce was washed away, and the soil was ruined by salt water. Virtually all of the equipment needed by fishing communities there was also destroyed. After the event, the inhabitants abandoned the island and settled elsewhere on the North Carolina coast.

There has not been devastation as drastic since then, although there have been several times, such as during Isabel in 2003, when storms have opened new inlets, at least for a few hours or a few days. But increased development makes the coast extremely vulnerable, and there is nothing in meteorological theory that suggests that storms of this magnitude will not come again.

INLAND TRACK

Although many fewer storms take inland tracks, they can bring the same wind and rain danger as those that stay on the coast. Hugo took this track in 1989 (see Fig. 5.5). Hugo came inland as a category 4 storm near Charleston. It lost much of its initial force over the South Carolina low country and reached Charlotte as a tropical storm. It continued in this form as it moved through our western Piedmont and eventually decayed near Lake Erie. In North Carolina there was a major storm surge on the Brunswick County beaches. Inland, much of the damage was done by winds, which reached 69 mph (gusting to 87 mph) in Charlotte; Hickory and Greensboro reported gusts of 81 and 53 mph, respectively. Falling trees did a tremendous amount of damage throughout the track, and thousands of dwellings were destroyed and damaged in Charlotte alone. In addition, downed trees made clean up and power restoration extremely difficult. Timber loss was also significant. Rainfall was fairly light; most of the Piedmont reported less than 3", although Boone measured 6.91". This, for us, was a billion-dollar storm, and seven people in North Carolina lost their lives.

Just to emphasize that inland storms can take a variety of often surprising tracks, we can look at the storm of August 1940. On August 11 it made landfall near Savannah and went west across Georgia, north through Tennessee, and into Virginia (see Fig. 5.5). There was nothing unusual about this track; it was somewhat like Hugo's but farther south and west. Indeed, we would expect the storm to carry on northward and die out, like Hugo, near the Great Lakes. But in this case, on August 14 it turned sharply and entered North Carolina from the northwest. It then followed a curving route over the state before leaving the northeast on August 17. Mostly it was a tropical storm without high winds, but all of the state got rainfall; more than 15" fell in much of the north. Floods were widespread, and landslides were common in the mountains. The East Tennessee and Western North Carolina Railroad line was so badly damaged that it ceased operation; it survives only as Tweetsie Railroad in Boone. The hurricane track, however, indicates that nowhere in our state is immune from hurricanes.

GULF STORMS

Hurricanes that hit the coast of the Gulf of Mexico only rarely make it into our state. Usually they are rather weak remnants, often move slowly, and sometimes bring lots of rain. To penetrate this far north they have to be fairly intense to start with. Hurricane Camille in 1969 took this track but just missed us. Camille hit the gulf as a category 5 storm, one of only two such storms to hit our nation in the twentieth century. The storm moved northward through Tennessee and then swung eastward to do major rain, wind, and flood damage in the Virginia and Washington, D.C., areas. We were not so lucky in 1916. A storm came north from the gulf, slowing and weakening as it moved. It stalled over Tennessee and gave tremendous rainfall over east Tennessee and western North Carolina. This was the second storm in a month, and major flooding resulted (see Box 5.3).

FIGURE 5.5. *Tracks of hurricane Hugo in 1989 and the hurricane of August 1940. Solid lines indicate the hurricane stage; dashed lines, a tropical storm; a dotted line, an extratropical storm.*

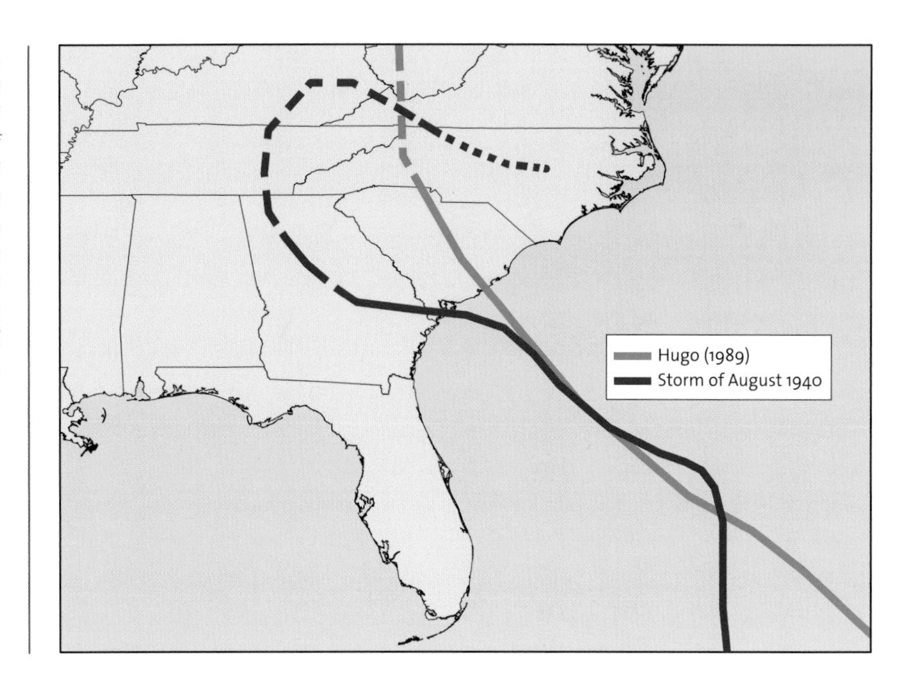

YEARS WITH MULTIPLE HURRICANES

It is common for North Carolina to be influenced by more than one hurricane in a single season. Hazel was the third of three storms in 1954; the first two tracked off our coast and caused only storm surge damage. The following year there were also three hurricanes. The first two, a category 3 and a category 2, crossed the eastern part of the state within a week of each other. They caused much wind- and rain-related damage, including some flooding. A month later came another category 3 storm, Ione, which also followed the basic coastal track and produced even more rain. After back-to-back multihurricane years in 1954 and 1955, the next such year did not come until 1996. Even then, the main storms were separated by two months and had different tracks. The first, Bertha, was the first category 2 or greater storm to track inland through the Coastal Plain since Diana of 1984, and Bertha was also the first to have a major inland impact since Donna in 1960. There was a lot of damage, mainly wind related, in an area where there had been a great increase in development since the last hurricanes. Bertha also provided much rainfall, which had a major influence when hurricane Fran came along a couple of months later. After the two hurricanes of 1996, we only had to wait until 1999 to get another multihurricane year that brought Dennis and Floyd (see Box 1.2).

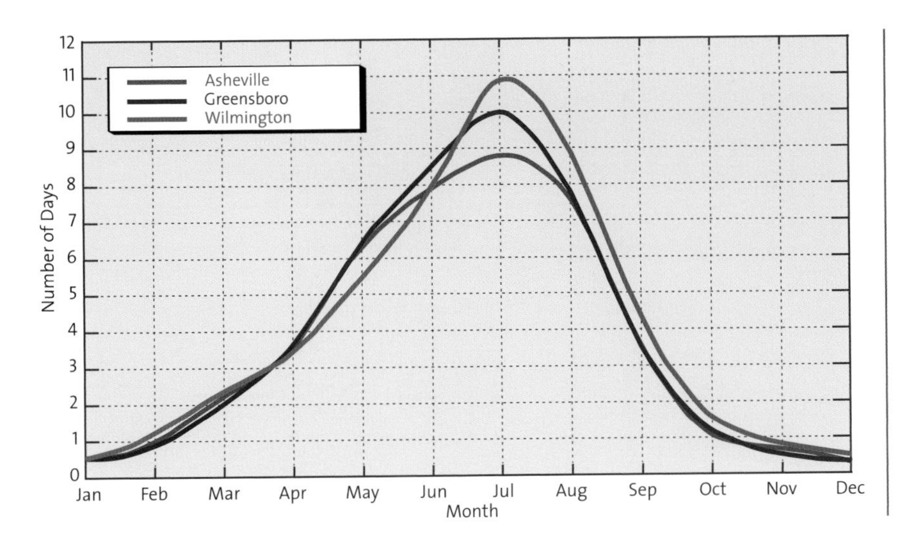

FIGURE 5.6.
Average number of days per month with thunderstorm activity for three North Carolina stations. July is the most active month, but winter thunder occurs in most years.

Thunderstorms and Tornadoes

Thunderstorms can bring intense rain, hail, lightning, and tornadoes. In North Carolina they can occur at any time of day or night and at any time of year (see Fig. 5.6). Nevertheless, the most likely time for a storm is the late afternoon in summer. The mountains get slightly fewer days with thunder than does the Piedmont, while the Coastal Plain gets more than either. But the difference is very small, and the whole state gets about forty-five days with thunder in an average year.

According to the NWS, a thunderstorm starts when the first thunderclap is heard and ends fifteen minutes after the last peal. If thunderclaps are more than fifteen minutes apart, a new storm is registered. In our part of the world the noise of thunder travels about five miles. So a thunderstorm at a place is really an event that occurs within about a five-mile radius. (In the American West, that distance may be twelve miles.)

Most days have only a single storm, but some days have a series of storms occurring one after the other. Multiple storms are most common around the Sandhills and in the mountains, where the storms really do seem to roll around in the hills. So the number of storms we get per year varies from about fifty in the northeast to sixty in the mountains and approaches seventy in the south. Nationally, that puts us in the middle: much of the midsection of the nation has values near ours, with a general decrease in numbers northward. The West has far fewer. The gulf coast and Florida have many more than we do; the Tampa Bay area has twice as many and proudly proclaims itself the thunderstorm capital of the nation.

FIGURE 5.7. *Life cycle of a thundercloud, expressed as three sequential stages*

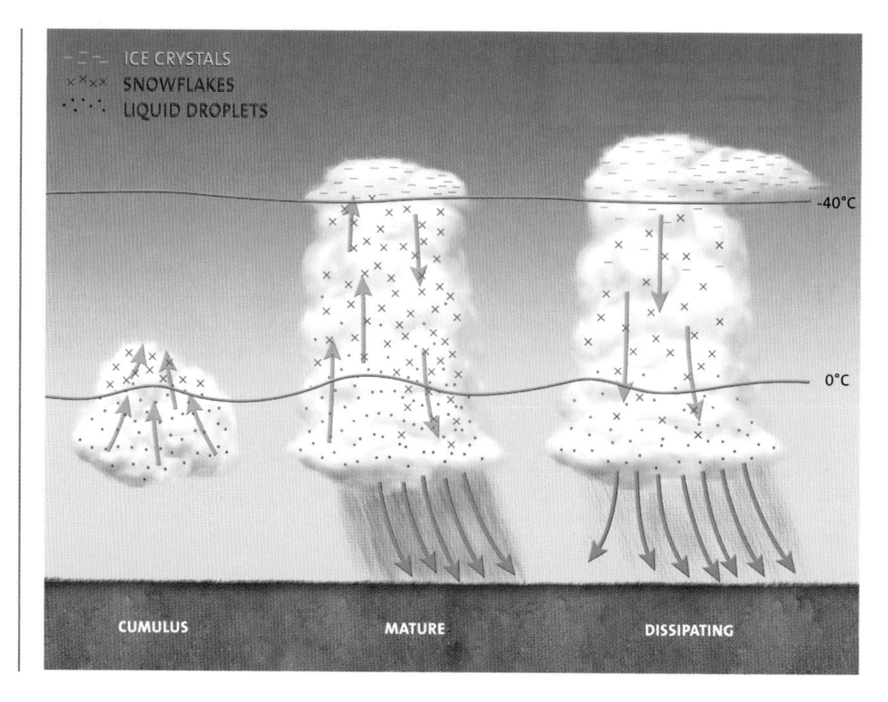

THE LIFE OF A THUNDERSTORM

It is often possible to observe the entire life of a thundercloud (see Fig. 5.7). Typically we start our summer days with a cloudless morning. In the afternoon small, fluffy cumulus clouds start to appear. Some may slowly build from the incipient stage into the cauliflower-like tower of a mature thundercloud, which tilts slightly forward and drifts along with any upper wind. We may see an "anvil" at the top, leading the cloud. This indicates the *tropopause* level, which acts as a lid for the ascent, with the air blown forward by the strong winds there. At that level the cloud water is frozen, and ice crystals make up the wispy anvil shape. If the cloud passes over us at this stage, we shall get a sequence of weather. First comes a gusty period of cold air as the gust front reaches us. Temperatures fall because we are in the downdraft air. This becomes obvious because the rain, starting with a few scattered large drops but soon becoming an intense downpour, arrives. As the cloud moves by, the rain decreases in amount and intensity and eventually ceases. The cloud has passed.

The driving force for our cloud has been the instability created by the very warm surface air, but the downdraft brings in colder air that cools the surface. This decreases the instability; soon there is not enough energy to drive the updraft, and so the cloud dissipates. The whole life cycle of an individual cloud may take about thirty minutes. But one thundercloud may spawn others. The gust front pushing ahead of the mature cloud often undercuts, rather than mixes with, the warm surface air. This encourages the development of a new cloud, or even several clouds, ahead of the old one. The new clouds then go through their life cycle, perhaps also spawning new clouds. The result is the thunderstorm, composed of numerous individual clouds, perhaps lasting several hours, and traveling many miles.

The thunderstorms described above, driven entirely by local instability, are *air mass storms*. They can occur at any time when the maritime tropical air mass is overhead, but they are most common on a summer afternoon. We get a series of isolated thunderstorms scattered around the state in a seemingly random fashion. They may give plenty of rain, sometimes intense but usually highly localized, which may cause some very local flooding. But they rarely lead to truly severe weather. Occasionally we get a line of well-organized thunderstorms moving over the state, often associated with the passage of a cold front, where the convection is much more vigorous and severe weather is more likely. All storms, however, produce thunder and lightning, and may bring hail.

THUNDER AND LIGHTNING

Thunder and lightning result from the separation of electrical charge caused when water molecules rub together within the thundercloud. A positive charge develops near the cloud base, and a negative charge forms near the top. Once the potential difference between them has built enough to overcome the resistance of the air—which may involve millions of volts—a lightning stroke occurs. Each stroke is made up of many individual flashes, sometimes in several directions, in a brief interval. The strokes may be within one cloud, from cloud to cloud, or from cloud to ground. Whichever occurs, the almost instantaneous passage of an electric current through the air very rapidly heats that air. It expands "explosively," causing a pressure wave to move outward, and our ears respond to that pressure change. We hear thunder. The pressure wave is soon damped by the atmosphere, and the audibility of the thunder is limited to roughly five miles. The light of lightning, however, travels just like sunlight. During the day we rarely notice distant, seemingly silent lightning because it is swamped by ordinary daylight. At night we are much more aware of the distant thunderstorm, since we can see the apparently noiseless "summer lightning."

Cloud-to-ground lightning is of most practical concern for us. In this form, a leader stroke moves down from the cloud base. When the leader is close to the ground, an upward stroke from a nearby upstanding point connects to the leader, completes the circuit, and transfers the electrical charge to the ground. While the upstanding point is not always the highest point in the area, it often is, and in most lightning-protection schemes a grounded metal lightning rod is made the local highest point. Whenever this connects with the downward leader, the current is conducted through the rod and grounding lines into the earth, reducing the chance of the current flowing through and damaging the protected structure.

People, of course, can provide the upstanding point for lightning, and injuries or death can result (see Table 5.5). Nowhere outdoors seems safe, and golf courses are no more dangerous than open fields. Sheltering under trees is never a good idea. Indoors, including in an automobile, is the safest place to be in a thunderstorm.

HAIL

Within the turbulent air of a growing thundercloud, the individual droplets move about at random. Eventually they become better organized, and an almost circular motion may start. During each circuit the droplet grows as moisture condenses around it. In many clouds the circuit will move the drop above and below the freezing level. During the time that it is below, the outer layer of the drop will be liquid and have a more or less spherical surface. When it is carried above the freezing level, that liquid layer will freeze, and more ice will be added. When it falls again, a thin surface layer may melt, but more liquid water will be added to form another layer on top of the frozen core. When the cycle is repeated several times, a hailstone with an onionlike layering of ice is formed. When it grows large enough to overcome the upward motion of the cloud, we get hail at the ground.

Observations of hailstorms come mainly from the official reports of law enforcement, emergency management, or similar personnel; from newspaper reports; or from insurance company records. The big events are usually well recorded, and the smaller ones that do little damage are probably not well observed; thus the assessment of long-term trends is difficult. However, we can make some reasonably consistent estimates for recent years (see Table 5.6). The early 1990s seem to have been a fairly quiet time for hail, but the end of the decade was much less so. The amount of damage varies tremendously and depends greatly on where the storm hits. Of the property damage in 1997, for example, $6 million was the result of a single storm that virtually destroyed a car sales establishment in Jacksonville.

TABLE 5.5. Lightning Deaths and Injuries in North Carolina, 1959–1999

	OPEN FIELDS AND SPACES, BALLPARKS	UNDER TREES	WATER-RELATED ACTIVITIES	WORK; NEAR MACHINERY	ON GOLF COURSES	AT TELEPHONE	OTHER	NATIONAL RANK
Deaths	41	25	23	6	8	7	71	3
Injuries	161	43	28	45	28	11	261	4

Note: The top five states for lightning deaths are Florida, Texas, North Carolina, New York, and Ohio; for injuries, Florida, Pennsylvania, Michigan, North Carolina, and New York.

Fortunately, we do know that for the last few decades there have been no records of human deaths or major injuries from hail.

TORNADOES

Tornadoes are probably the most violent and unpredictable storms we have. They are much smaller than hurricanes, but they pack stronger winds and move faster and more erratically. They form below a thundercloud and move with the cloud. Sometimes they will remain above the ground for part of their track and do little damage. At other times they may touch down and cause major destruction. The wind speed within them, and the damage caused, also varies tremendously. The Fujita scale is used to classify tornado intensity (Table 5.7).

Fortunately for North Carolina, major outbreaks of tornadoes above F1 on the Fujita scale are rare, although not unknown. Certainly we cannot compete with the American Midwest in the intensity and number of tornadoes. This is almost certainly because our common air mass thunderstorms do not have the great amount of energy, and our atmosphere lacks the particular layered structure, needed to generate the severe storms found there. However, how a tornado develops from a thunderstorm is far from clear and is the subject of a great deal of research. The increasing amount of information about the interior workings of the storms available from Doppler radars is helping tremendously. Currently we think that the key to tornado formation is the development of a circular motion within the ascending air at the backside of the storm (see Fig. 5.8). This may even start as a horizontal rotor cloud that becomes tilted upright. This circular wind does not allow much air to penetrate into the center. As long as the central ascending air can be removed by the airflow aloft, the pressure, already low, falls further. This, in turn, speeds up the wind, which begins to suck in air from the

TABLE 5.6. Number of Hailstorms and Damage Estimates in North Carolina, 1993–2000

YEAR	NUMBER OF STORMS	PROPERTY DAMAGE ($K)	CROP DAMAGE ($K)
2000	347	620	1,981
1999	154	175	225
1998	547	2,685	2,003
1997	242	6,339	17,650
1996	254	14,267	840
1995	159	6,187	2,066
1994	51	0	5,005
1993	106	75	0

surrounding regions. This squeezes the center, the low pressure area gets smaller, and the winds automatically get faster—in the same way that a whirling ice skater draws in his or her arms when he or she wants to speed up. Condensation and the development of the dreaded funnel cloud usually build downward from the base of the thundercloud, although in some cases they appear to build from the bottom up. It may be both; we are not at all sure. In any event, we now have a tornado. This tracks along with the parent thundercloud. It may last for many minutes, although usually less than an hour, and cover many miles.

Within North Carolina it has long been assumed that the Sandhills region was most favorable for the development of thunderstorms and tornadoes. The area certainly has a high incidence of tornadoes, which partly justifies its reputation as our "tornado alley" (see Fig. 5.9). It is suspected that the high surface temperatures in the region foster the great instability and strong updrafts necessary. Furthermore, it is suggested that tornadoes formed in the Sandhills are moved by the prevailing southwest winds toward the northeast, where there are also areas of high incidence. But several Coastal Plain counties, and even some in the Piedmont, which are well away from the Sandhills, also have a high incidence of tornadoes. Recent research suggests that an additional factor is important: the energy contrasts between wet and dry surfaces that may foster the circular motions near the surface required for tornado formation. The northern counties of our state, and most of those in the mountains, are away from these favorable spots, and tornado incidence is low there. But few counties are immune.

Although our tornadoes can rarely compete in intensity with those of the Midwest, severe outbreaks are not unknown. One such outbreak, associated with the

TABLE 5.7. Fujita Scale for Tornadoes

SCALE	WIND ESTIMATE (MPH)	TYPICAL DAMAGE
F0	< 73	Light Damage: Some damage to chimneys; branches broken off trees; shallow-rooted trees pushed over; sign boards damaged.
F1	73–112	Moderate Damage: Peels surface off roofs; mobile homes pushed off foundations or overturned; moving autos blown off roads.
F2	113–57	Considerable Damage: Roofs torn off frame houses; mobile homes demolished; boxcars overturned; large trees snapped or uprooted; light-object missiles generated; cars lifted off ground.
F3	158–206	Severe Damage: Roofs and some walls torn off well-constructed houses; trains overturned; most trees in forest uprooted; heavy cars lifted off the ground and thrown.
F4	207–60	Devastating Damage: Well-constructed houses leveled; structures with weak foundations blown away some distance; cars thrown and some heavy missiles generated.
F5	261–318	Incredible Damage: Strong frame houses leveled off foundations and swept away; automobile-sized missiles fly through the air and are carried more than 100 yards; trees debarked; incredible phenomena will occur.

passage of a sharp cold front, occurred on March 28, 1984, and produced twenty-five tornadoes that caused forty-two deaths (Box 5.2).

Waterspouts are tornadoes that occur over water as they suck surface water into their circulation and thus become visible. On a much smaller scale, "dust devils" are tornado-like features that can occur almost anywhere in our state in summer. The intense heating of the ground, particularly if it is bare soil so that all of the sun's energy goes directly into heating the surface rather than evaporating or transpiring water, creates strong low-level instability. In calm conditions the resultant vertical motion can be given a twist—for reasons that are not understood—and a small rotor is created. If this picks up dust, we see it; if it does not, we don't. We do not know how many of these occur completely unobserved.

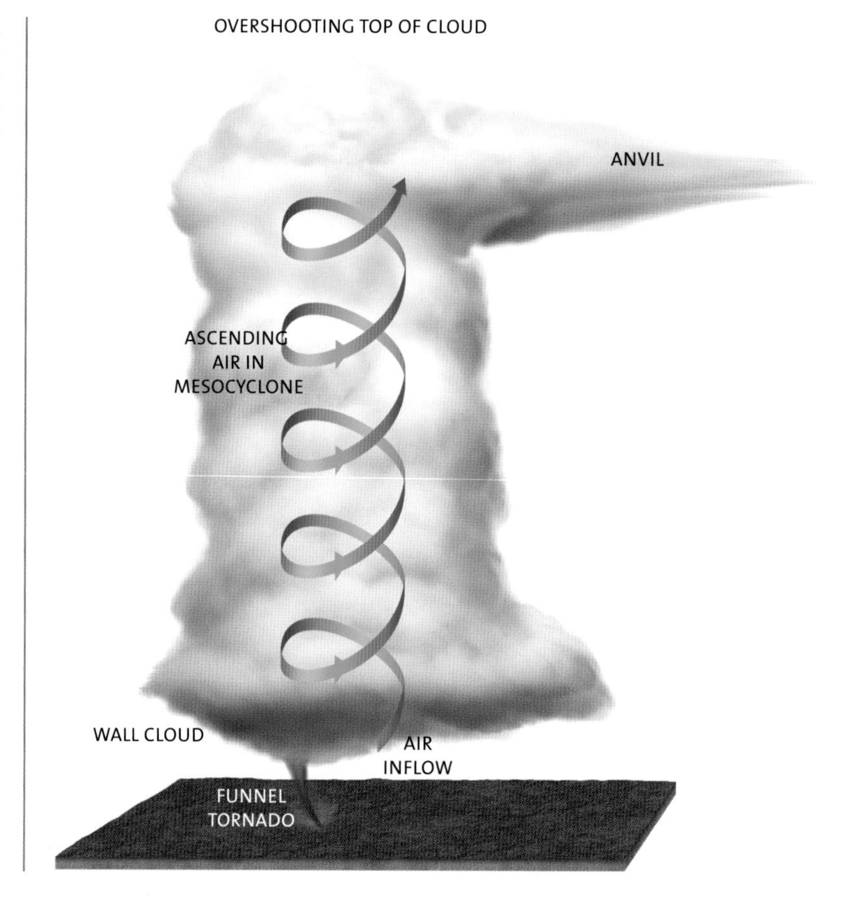

FIGURE 5.8. *Major features of a tornado and its associated thundercloud*

Floods and Droughts

Floods and droughts are both direct results of rainfall, or the lack of it. Floods usually come and go within a matter of hours or days, affecting rather small areas of the state. Droughts build silently for weeks, may last for months, and almost always cover a large area. While it is useful to consider the two phenomena together as impacts of rainfall, they differ in their causes and in how we can forecast and respond to them.

TYPES AND CAUSES OF FLOODS

Floods associated with rivers have a continuous range in size, from those affecting a single, small stream to events covering several thousands of square miles and

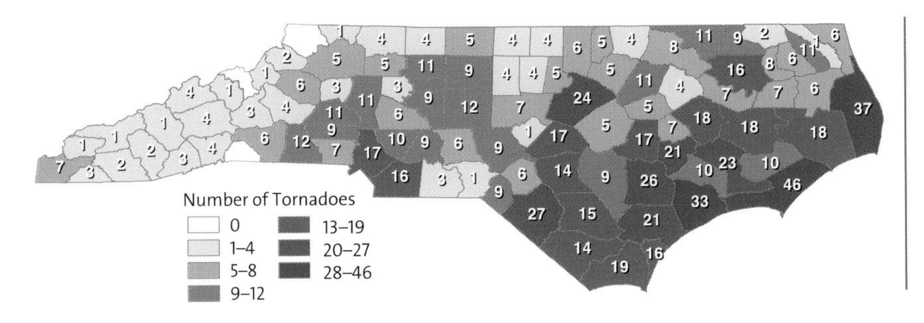

FIGURE 5.9. *Distribution of tornadoes by county in North Carolina, 1950–2003*

involving numerous rivers and their tributaries. Meteorologically, it is useful to separate floods into two types: (1) *flash floods*, which are local floods that occur very rapidly from intense rainstorms and are difficult to forecast more than a few minutes in advance, and (2) *widespread river floods*, which cover a wide area, build over several hours, and can be forecast far enough in advance to minimize loss of life, if not of property.

A flash flood is possible whenever there is an intense downpour onto a local area. However, a downpour alone is not likely to cause a flood unless other factors are favorable (see Fig. 5.10). These factors can be summarized as the things that impede percolation of rainwater into the soil, that encourage surface runoff, or both. Most aspects of the development of human settlements and their associated infrastructure do both, so urban areas tend to be more susceptible to the creation of flash floods than do rural areas. Urban areas are also, of course, usually more susceptible to the effects of the floods, since development is all too common in the areas likely to be flooded. In addition, the transition from the gentle base flow of a river to the torrent of a flood is likely to be much faster in an urban area than in a rural one (see Fig. 3.18). This makes forecasting floods even more of a challenge.

Flash floods are commonly associated with convective storms. The individual storm clouds produce intense precipitation for a short period, usually over a restricted area. So they have an impact only when they fall on the catchment of a small stream. The small catchment concentrates all the runoff directly and rapidly into the stream, which floods during or immediately after the storm. The flood subsides almost as quickly as it arose. The impact on the larger downstream rivers is small, since rarely are more than a couple of the tributaries affected by the same storm system at the same time. Further, flash floods tend to be felt more strongly in the mountains and foothills than on the Coastal Plain, largely because of the steeper slopes and thinner soil. Such floods are rare in the Sandhills, where the sandy soils can rapidly absorb large quantities of water.

BOX 5.2. THE TORNADO OUTBREAK OF MARCH 28, 1984

The most severe outbreak of tornadoes ever recorded in North Carolina occurred on March 28, 1984. A squall line associated with a strong cold front developed over South Carolina and produced numerous thunderstorms and some tornadoes there. The line moved northeastward into North Carolina, with the major thunderstorm activity concentrated along a line between Charlotte and Lumberton. Our first reported tornado touched down in Union County at 5:10 P.M. local time. Twenty-five tornadoes later, at 9:20 P.M., the squall line system died as it moved over Chowan and Perquimans counties into Virginia. Most tornadoes followed a track from southwest to northeast, along with the parent squall line, but they varied tremendously in width, length, and intensity. Some were of an intensity rare in North Carolina and usually restricted to the Midwest, where they are categorized as "violent." In North Carolina the tornadoes between them did more than $325 million in damage, killed 40 citizens, and injured 400 others. In terms of the loss of human life, this was one of the greatest natural disasters the state has ever known.

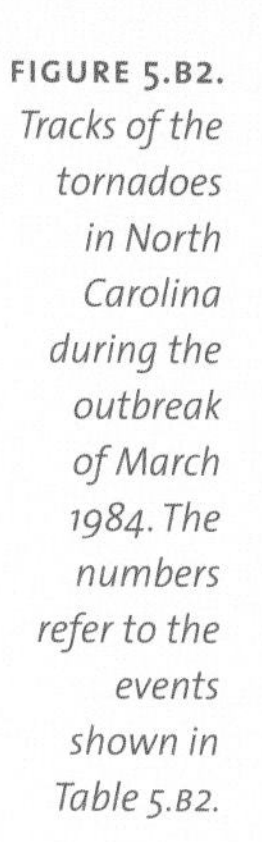

FIGURE 5.B2. *Tracks of the tornadoes in North Carolina during the outbreak of March 1984. The numbers refer to the events shown in Table 5.B2.*

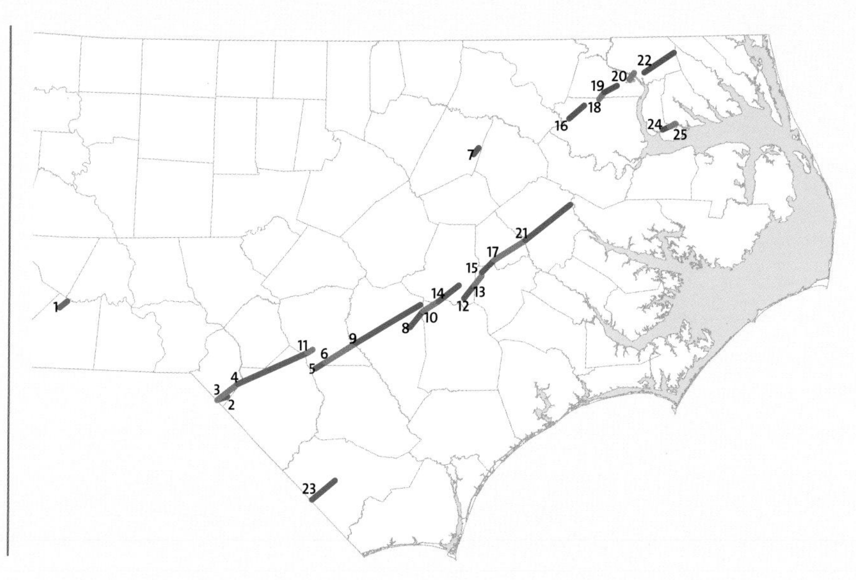

Statewide there is an annual cycle of flash flooding. Early in the year there are few thunderstorms to produce intense local rainfall. The flash floods start to occur as spring advances, when convection increasingly delivers intense rainfall onto saturated soil. This pattern reaches its climax in June. After that, the rapid

TABLE 5.B2. The Tornadoes of March 1984

STORM #	TIME	COUNTY	FUJITA	LENGTH (MILES)	WIDTH (YARDS)	DEATHS	INJURIES
1	1710	Union	1	0	33		
2	1825	Scotland	4	3	2,113		
3	1825	Scotland	4	8	2,640		
4	1840	Robeson	4	25	2,640	2	280
5	1845	Bladen	3	4	1,407		
6	1850	Cumberland	3	11	1,407	2	11
7	1910	Nash	2	2	177		
8	1915	Sampson	4	6	1,407		
9	1915	Sampson	3	25	1,407	10	90
10	1925	Duplin	4	7	1,407		40
11	1930	Cumberland	4	2	2,640		
12	1930	Wayne	3	5	527		
13	1937	Lenoir	3	4	527		74
14	1940	Wayne	4	8	1,407		7
15	1945	Lenoir	4	5	1,223	3	59
16	1955	Bertie	3	6	880		
17	1955	Greene	4	13	1,223	6	19
18	2010	Bertie	2	3	527	7	
19	2015	Hertford	2	2	527		2
20	2020	Hertford	1	1	177		5
21	2020	Pitt	4	20	1,223		
22	2030	Gates	3	14	880	9	153
23	2045	Columbus	2	9	353	2	10
24	2115	Chowan	2	4	177		
25	2120	Perquimans	2	2	177	1	1

evaporation of soil water in summer keeps the soil fairly dry, so that only the most intense systems are able to produce enough water to first fill the soil and then create the flood runoff. As the intensity of convection tapers off in the fall, so does the frequency of flash floods.

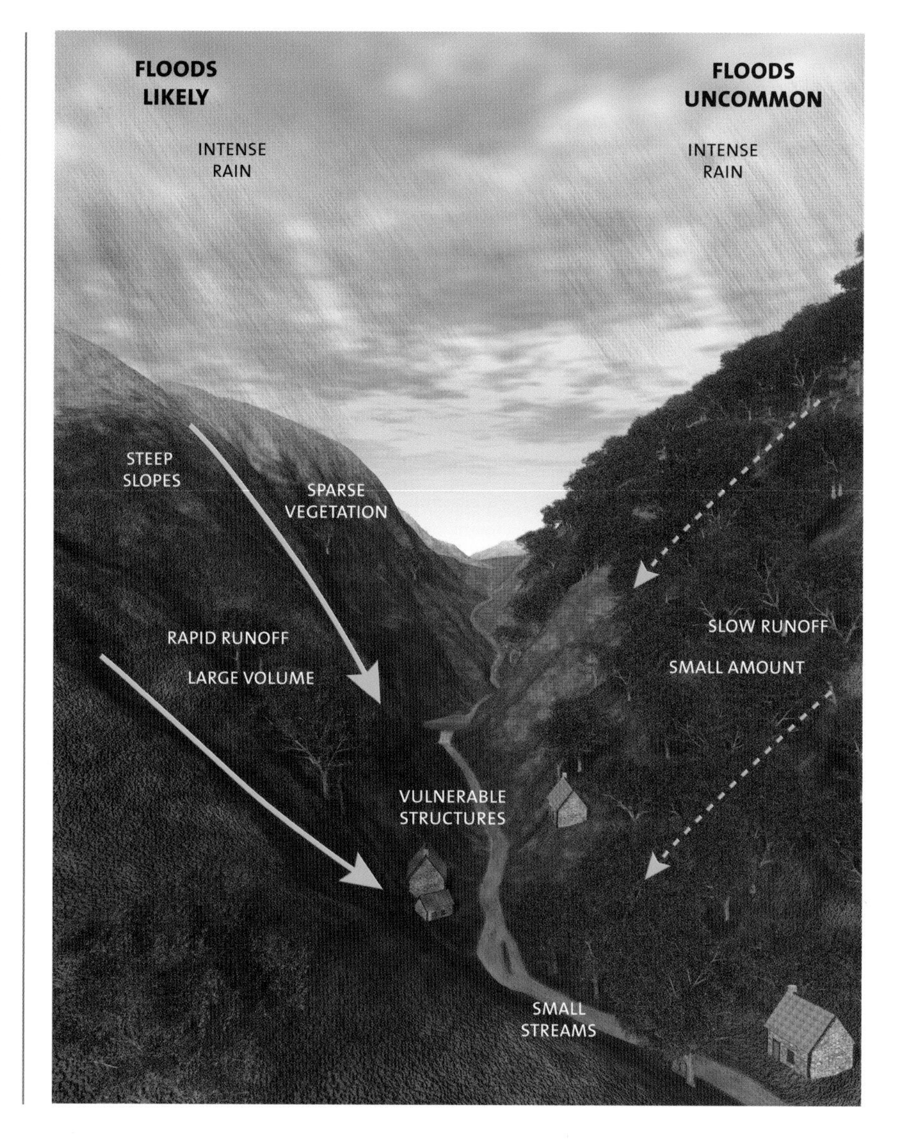

FIGURE 5.10. *Factors favorable for the creation of floods. The situation for small-scale flash floods is shown, but similar conditions are needed for widespread river flooding.*

Widespread river floods cover a range of sizes and impacts but are generally the result of a storm system bringing copious rain onto a soil that is already at or close to saturation. Often the widespread rain from a wave cyclone ensures that the soil is nearly full of water, then the frontal passage provides the extra, perhaps intense, rain needed to create the rapid runoff leading to the flood. Because a wide area is involved, it is most likely to be rural, and the influence of urbanization on the creation of these floods is not great. But again, humans often build in the floodplains of rivers likely to flood, so the impact of the floods increases with urbanization.

For most of the state, these widespread floods occur most often in late winter and early spring. At these times the soil is likely to be full of water, so a relatively small amount of rain from a passing cyclone or frontal system is needed to start a flood. In summer and fall, soils tend to be dry, much more rain is needed, and a single rainfall event is rarely enough to produce floods. But this is the season for hurricanes, with their great rainfall amounts. So our major floods, although infrequent, usually occur in the fall.

In the mountains in middle and late winter the soil over a wide area may become frozen. This encourages rapid runoff and hence some degree of flooding from virtually any rainstorm in the area. Thus winter is the major time for widespread mountain floods.

ONE HUNDRED YEARS OF MAJOR FLOODING

Like other events associated with the atmosphere, the frequency of flooding varied during the twentieth century (see Fig. 5.11). The major floods represent times when two or more of our large drainage basins had simultaneous flows so high that they were unlikely to occur more than once about every ten years. Severe floods, with flows unlikely more than once every twenty years, were also identified. Using these criteria, there were thirty-one such floods, ranging from one that affected the area between Charlotte and Asheville in May 1901 to the flood created by Dennis and Floyd in the east in September 1999. Of the seven severe floods, only two occurred in the last half-century, while very few floods, severe or major, occurred between 1931 and 1970. Part of the decrease after about 1930 is almost certainly the result of dam building and flood control on our major rivers. But the return of floods after 1970 probably has a meteorological cause.

When we look at the weather that created these floods, a clear pattern emerges. In almost all cases two weather events were necessary to create the flood. The first event made sure the ground was absolutely soaked and the rivers were already at least bank-full. The second system then dumped more water, and the flood resulted. Many of these dual weather systems were hurricanes. The floods of 1999, caused by Dennis and Floyd, are a case in point (see Box 1.2). But this event was not unusual, and major floods can occur over any part of the state. Those of 1916 were mainly in the mountains and associated with at least one gulf hurricane (see Box 5.3). Indeed, the frequency of major floods appears to be closely associated with the frequency of hurricanes (see Table 5.3). We did have a long period in the second half of the twentieth century without them. But the atmosphere is still capable of providing the tremendous volumes of precipitation required to create widespread floods. The possibility that hurricanes may become more common early in the twenty-first century suggests that we might also see an increase in the number of major or severe floods.

THE NATURE OF DROUGHT

In many ways drought—a prolonged, silent feature that creeps up on us insidiously and has simultaneous impacts on a large portion of the state—is the opposite of flooding. While drought obviously starts as a "day without rain," we have many such days—or even long sequences of dry days. But we normally think of drought as starting when we begin to feel the impacts of that lack of rainfall, whether it is the wilting of the crops in our fields and the plants in our yard, the falling water levels in our wells, or the low flow of rivers making fishing or boating difficult. So we define drought in terms of its impact rather than by a strict meteorological definition. Unfortunately, all the various impacts do not make themselves felt at the same time or in the same way. So there is no single kind of drought and no single definition. However, meteorologists have divided the impacts into two main classes and consider agricultural and hydrologic drought separately.

Agricultural drought is primarily connected with the impact of the lack of rainfall on crop growth, although it also directly affects the growth in our gardens or, indeed, all vegetation growth. As such, it involves the amount of soil water available to plants. In our climate, soil water amounts are usually lowest at the end of summer (see Chapter 3). Even with a rainy summer, there will be some periods where many plants will feel water stress. With less than normal rain, there may be prolonged periods when the soil reaches the wilting point and agricultural drought conditions prevail. During the following winter the soil will fill up, and at the end of most winters the soil is at field capacity. So agricultural drought is a problem of the growing season, associated with low summer precipitation, and in North Carolina it rarely persists from one year to the next.

Hydrologic drought is much more concerned with the water available for public water supplies. Rainfall provides the water. Some of the rainwater will run off to streams almost immediately, while some will be passed through the soil to become part of the natural underground reservoir. Slowly this groundwater is removed as it feeds springs and streams (see Fig. 5.12). Our extraction of well water is also a removal. During periods when precipitation exceeds removal, groundwater levels and stream flow increase. In dry periods, levels fall. Since the movement of

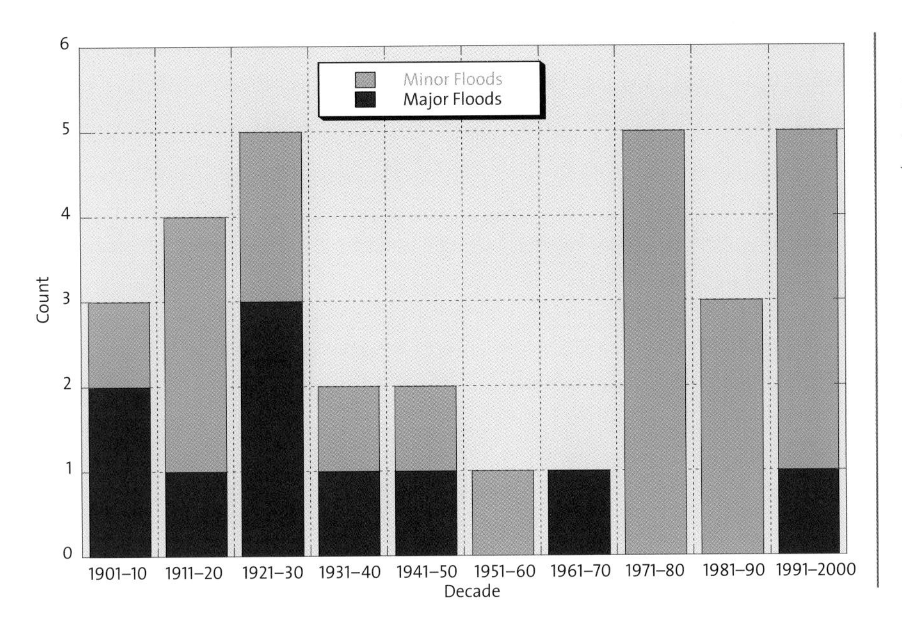

FIGURE 5.11. *Number of major and minor widespread inland floods per decade in North Carolina throughout the twentieth century*

water through the ground is slow, the impact of an individual rainstorm is hardly noticed. Dry periods that last a couple of months or more, however, will cause a noticeable drop in groundwater levels and stream flows and will herald the onset of a hydrologic drought. Unlike agricultural drought, hydrologic drought is not tied to the seasons, and the drought may last for several years.

Drought periods of either type may not be completely without rain. Any extended period with less than normal rainfall can create drought. Agricultural drought in North Carolina is often caused when the Bermuda high moves west from its normal position and sits over us. Then few convective storms can develop, while wave cyclones pass well north of us. The high can move over us at any time in the summer, and we simply have to wait for it to move away, which may be in a few days or a few months. Certainly, changes in the midlatitude circulation will ensure that it moves before winter. But one of the few systems that is almost certain to move it before then is a hurricane. There have been numerous years in which a hurricane has relieved our agricultural drought. The causes of hydrologic droughts are less distinct. Reduced rainfall, for example, may be produced by a seemingly slight and persistent displacement of the jet stream farther north than usual. This was the case with our 1950s droughts, when even clouds were rare. The droughts for the late 1990s appeared to occur during a period when vertical motions were less vigorous than normal. The clouds appeared, but they never gave rain (see Box 5.4).

BOX 5.3. THE MOUNTAIN FLOODS OF 1916

A major flood centered on the North Carolina mountains occurred in July 1916. This was associated with the remnants of two hurricanes that arrived in rapid succession. The first hit the gulf coast of Mississippi and Alabama on June 29. It soon lost force and became a tropical storm, meandering over the northern portions of those states for a few days. Eventually it swung eastward and then north, just grazing the western tip of our state on July 9–10 before dying out northwest of us (fig. 5.B3a). Throughout that time, the storm was producing heavy rain. Most streams with headwaters in the mountains were bank-full, and many were already in flood. Meanwhile, a second hurricane was coming inland, following much the same track that Hugo would take more than seventy years later. This was also a dying hurricane by the time it reached the western part of the state on July 14, but it produced copious rain, with amounts boosted by the orographic uplift of the air over the Appalachians. Many areas got more than 15" of rain during the forty-eight hours of the storm (fig. 5.B3b). Altapass, on the Mitchell/McDowell county line, had 22.22" in twenty-four hours, establishing a record for the greatest amount of precipitation in the United States during a twenty-four-hour period. This record stood for almost half a century until a station in Texas received even more. Rainfall from the two storms together also set records for monthly rainfall totals. Much of western North Carolina had more than 20" in the month; some places had 35" (fig. 5.B3c). *Annual* totals are usually around 40" in much of this area.

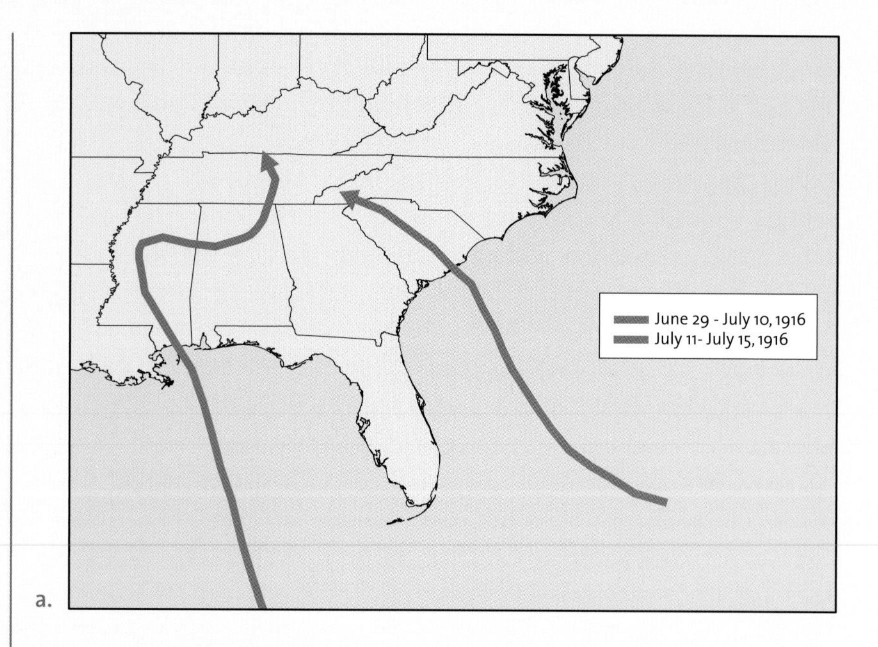

FIGURE 5.B3. *The great rainfall of August 1916, showing (a) the tracks of the two hurricanes that created the rain, (b) the forty-eight-hour rainfall totals from the hurricane passing through North Carolina, and (c) the monthly rainfall totals for July 1916 in North Carolina*

After the second storm, all the rivers in the western part of the state were in flood, and there was major destruction throughout the area. Forests had recently been cleared from many valley slopes, and orchards had been planted. Landslides rapidly moved loose soil, loose trees, and anything else in their path into already swollen rivers. The debris collected behind bridges, created dams, and increased the flooding until the bridges themselves were swept away. The Southern Railway lost at least half of its bridges on the lines radiating from Asheville, and in innumerable places the lines were blocked by debris. Damage was estimated at $22 million (in 1916 dollars), while at least eighty people were killed as a direct result of the flood.

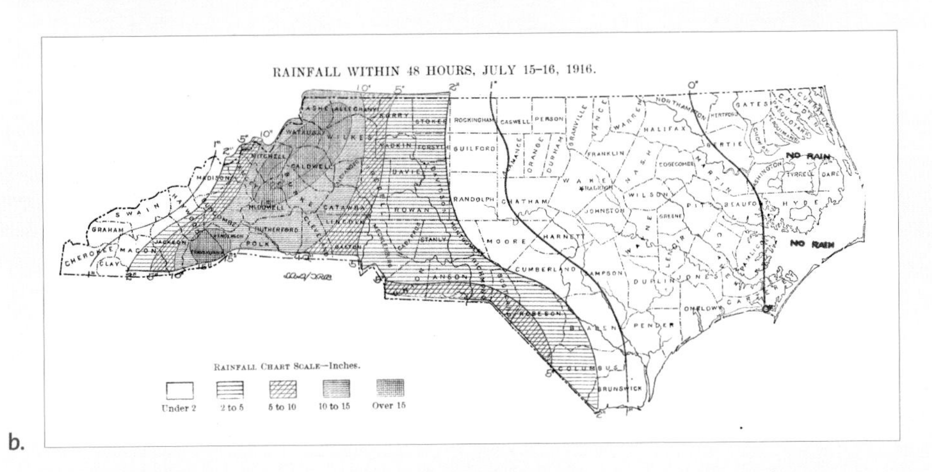

b.

c.

FIGURE 5.12. *Schematic of subsurface water flow to streams in normal (left) and dry (right) years*

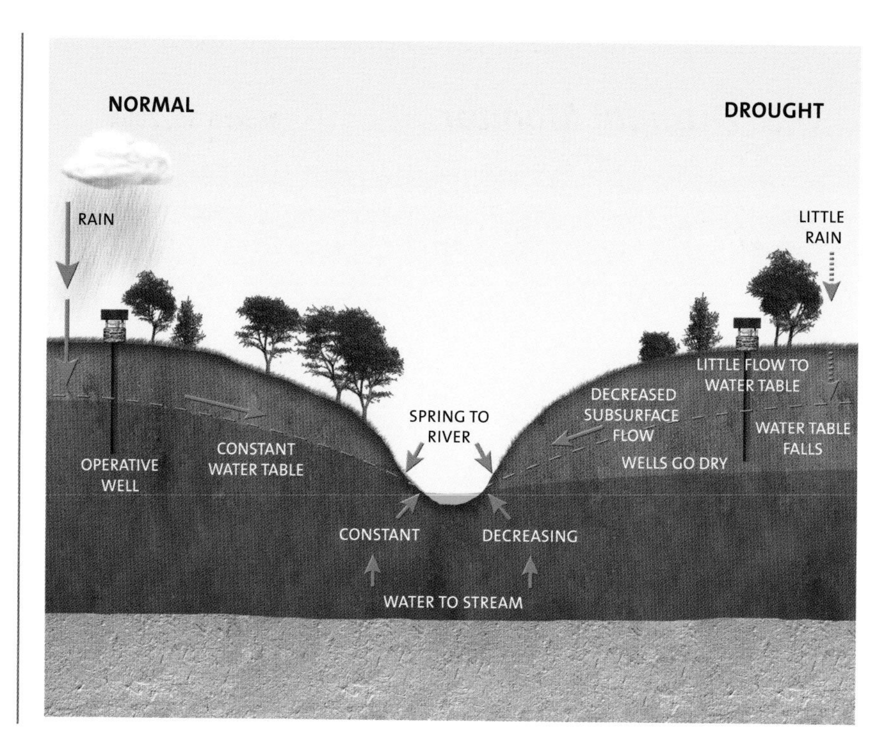

THE SEVERITY OF DROUGHT

Determination of the severity of a drought is vital, since a severe drought is as much a natural disaster as a severe flood, and those affected are equally in need of outside assistance. But determining severity is difficult because drought means different things to different people. Furthermore, drought conditions in North Carolina would be seen as abundant rainfall in arid Arizona. Several complementary measures are used to provide a drought intensity value (see Fig. 5.13). This does not give a "drought forecast" in the usual sense. Rather, it tracks previous conditions to determine the current drought severity. Thus it is possible to use historical weather records and long-term forecasts to suggest how long the drought is likely to persist and to determine the amount of rainfall needed to overcome the drought. Since a drought develops and is relieved slowly, communities have an opportunity to plan and implement measures to mitigate the hardship caused by drought. Most states, including North Carolina, have a drought monitoring body charged with developing and implementing such measures when warranted.

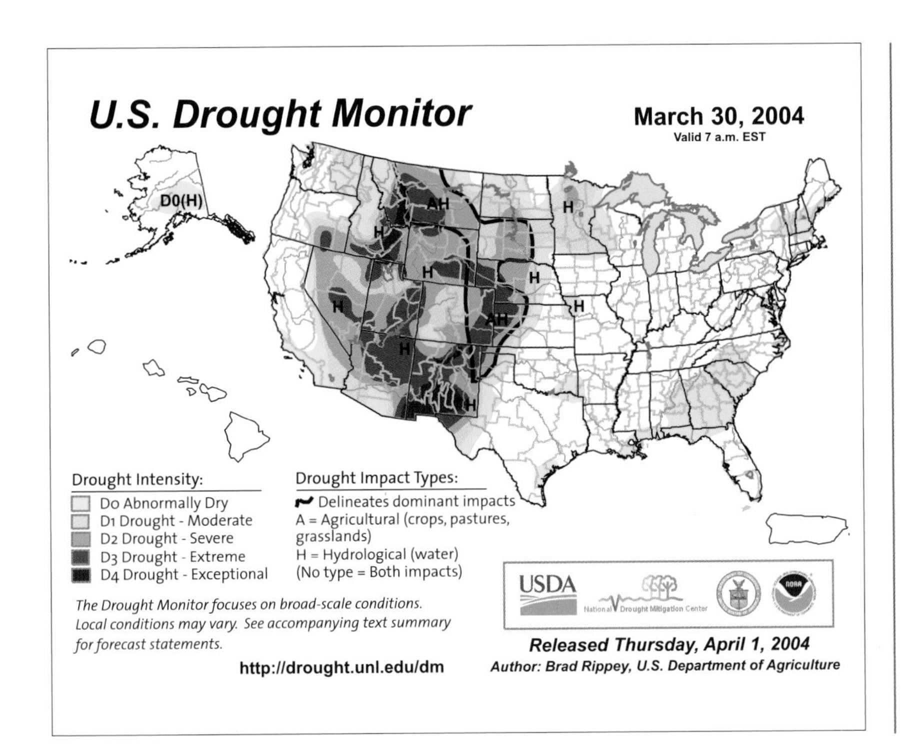

FIGURE 5.13. *An example of the analysis of the drought conditions on March 30, 2004. Meteorologically, North Carolina had no water shortage at this time, but water was in shorter supply than usual in much of the West. Our understanding of drought restricts the information we can give to monitoring, rather than forecasting, the progress of a drought. (image courtesy of the National Drought Mitigation Center, Lincoln, Neb.)*

DROUGHT IN TWENTIETH-CENTURY NORTH CAROLINA

During the twentieth century there have been periods when droughts were common and times when they were rare. They were uncommon until the mid-1920s. Then there was a ten-year period with very frequent droughts, some of which persisted for a couple of years and covered the whole state. This was the time of our most severe droughts. They created considerable economic problems in a state that was mainly agricultural, and these problems were made much worse by national economic conditions. The droughts ended in the wet winter of 1933–34, when the Dust Bowl period in the Midwest was getting under way. As the century progressed, there were some isolated and rather mild drought periods in the early 1940s and mid-1950s. Not until the 1980s was there another period when drought was common. Even then, droughts that affected the entire state tended to be short lived. In the 1990s it seemed that drought returned, with the mountains being the area most severely affected (see Box 5.4).

BOX 5.4. THE TURN-OF-THE-MILLENNIUM DROUGHT

At the turn of the millennium most of North Carolina was afflicted by drought, which was most persistent in the mountains. A long period of below-normal rainfall began there in the summer of 1998. The area involved expanded to include parts of the Piedmont the following winter and parts of the Coastal Plain by the summer of 1999. Only in the central Piedmont was it, for a few months, a severe drought, and the rains of Dennis and Floyd easily broke it there. These rains turned drought to flood on the Coastal Plain. Meanwhile, in the mountains the drought continued, slowly getting worse. In October 2000 the Palmer Drought Severity Index (PDSI)—the drought index most used in our area—fell below -3, signaling the onset of a severe drought. Stream flow was extremely low, groundwater was becoming scarce, and there was a growing need to reduce water use. This condition remained almost unchanged for about a year. Starting near the end of 2001 the drought area expanded to include virtually the whole state. By the summer of 2002 much of North Carolina was in extreme drought (PDSI below -4 and precipitation for the preceding six months less than 60 percent of normal). Virtually every community had problems obtaining water for human and agricultural use. In September 2002 abundant rain finally came to the mountains after more than four years of low rainfall and water problems. The Coastal Plain was also wet at this time, but over the Piedmont the extreme drought persisted for another month before abundant rain brought relief from the water problems. Direct costs of the drought amounted to tens of millions of dollars in lost agricultural production alone. The indirect costs associated with factors ranging from cities having to import water to individual buildings needing repair because of excess settling in the unusually dry soil are almost impossible to calculate. The North Carolina Rural Economic Development Center made a preliminary estimate that the drought cost at least $1 billion in damage, extra expense, and lost revenue.

This drought was one of the three worst in the last 100 years. A drought in 1986 was of shorter duration but even more intense. It started in the southern mountains early in the year, expanded to affect the whole state between May and October, and then subsided rather rapidly. Between June

Air Pollution

Air pollution affects North Carolina in many ways and in a variety of areas. Usually we think of dead trees on Mt. Mitchell and of ozone alerts in our major cities. But we have a range of problems, all of which share the same basic characteristics (see Fig. 5.14).

Human activities, rural and urban, put pollutants into the atmosphere, and the wind moves them away from their source. Different activities add different materials. Some are fairly large particles that do not rise very far and fall out and September the southern mountains had a PDSI of -5 (an exceptional drought in which rainfall in the preceding twelve months is no more than 65 percent of normal). July 1986 was the most intense drought month of the century, with the whole of the southern mountains and the Piedmont below -5. But probably the severest drought in our record was that between September 1926 and June 1927. All of the state outside the mountains had a PDSI below -3 in this period, with much of the Coastal Plain below -4. In the southeastern portion of the state the severe drought persisted almost until the end of 1927. And this drought came hard on the heels of one that lasted from July 1925 to January 1926.

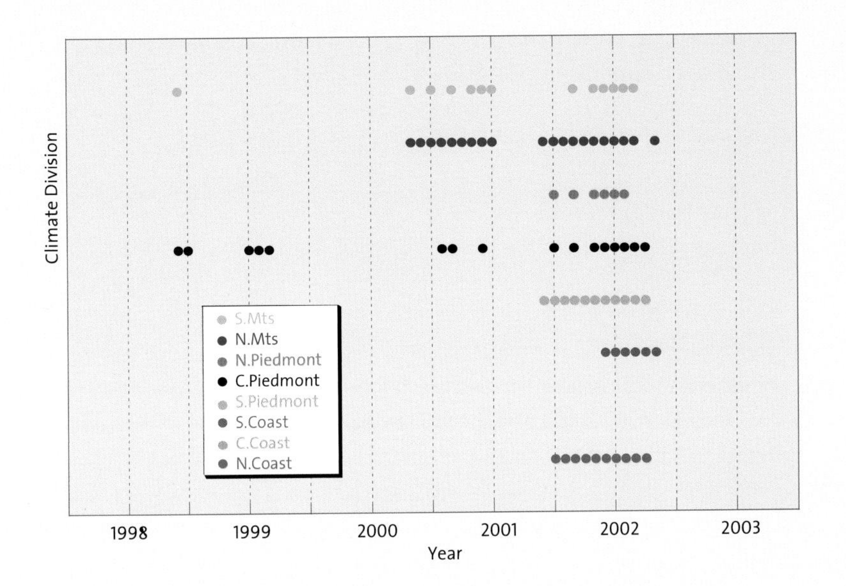

FIGURE 5.B4. *Weeks with PDSI less than -3 for each North Carolina climate division for the 1998–2002 period, centered on the turn-of-the-millennium drought*

close to their source. Others are smaller particles that can travel far downwind before they fall back to earth. Still other materials are gases that may mix with the atmospheric gases and be transported all around the earth. Whatever the type and source, all may interact with sunlight, with other gases and particles in the atmosphere, or with one another to produce new pollutants. All will eventually fall to earth directly or be incorporated into clouds and raindrops and washed to earth.

On a national scale, North Carolina is not a leader in pollution emissions. Nevertheless, an array of different pollutants are put into the atmosphere as a result of our diverse economy. For many pollutants the urban counties, mainly in the Piedmont, are the major source region. For other pollutants, individual industrial or power-generating sites, often in more rural areas, account for most emissions. Some pollutants are also blown into our state from neighboring areas, but we in turn export some of our pollution to others. In general, therefore, there is likely to be some kind of air pollution everywhere in our state virtually all the time. Fortunately, most of the time pollution levels are low because either the winds mix the near-surface polluted air with cleaner air aloft and carry the pollutants away or rain washes them out. But some meteorological situations are far from favorable, and high pollution concentrations, with their implications for health and other problems, can develop.

INVERSIONS, LOCAL POLLUTION, AND TROPOSPHERIC OZONE

Temperature inversions—when atmospheric temperatures increase with height rather than decrease as they usually do—prevent vertical air motions. They are often associated with low wind speeds, so pollution cannot disperse and dilute in the usual way. Any pollution material emitted stays close to the source, and concentrations increase. A day with a persistent inversion is likely to be a day with high pollution concentrations. Indeed, forecasting the onset and persistence of inversions is one of the major tasks for the air pollution meteorologist. Unfortunately North Carolina has a high frequency of inversions (see Fig. 5.15). They can occur at any time of the year and are common early in the morning following a calm, clear night. But usually they do not last for more than a few hours and are not very strong. Inversions that last for a few hours may occur in winter when a stream of warm air from the south blows over our cold land surface. But these kinds of inversions can occur almost anywhere in the eastern United States. The major reason for the high frequency in our area is the presence of the Bermuda high pressure region. The descending air within it is very warm and traps a layer of slightly cooler air at the surface under it, causing the inversion. The system also has low wind speeds, which ensure that trapped pollution cannot disperse horizontally. The Bermuda high often sits over our state for several days or even weeks in summer, allowing pollution concentrations to build over a long period of time.

The term "air pollution" covers a range of substances having a variety of possible health consequences. The U.S. Environmental Protection Agency has identified a group of criteria pollutants of particular importance that can be monitored relatively easily on a routine basis (see Table 5.8). The agency has also identified the concentration levels above which problems are likely to occur. Two thresholds

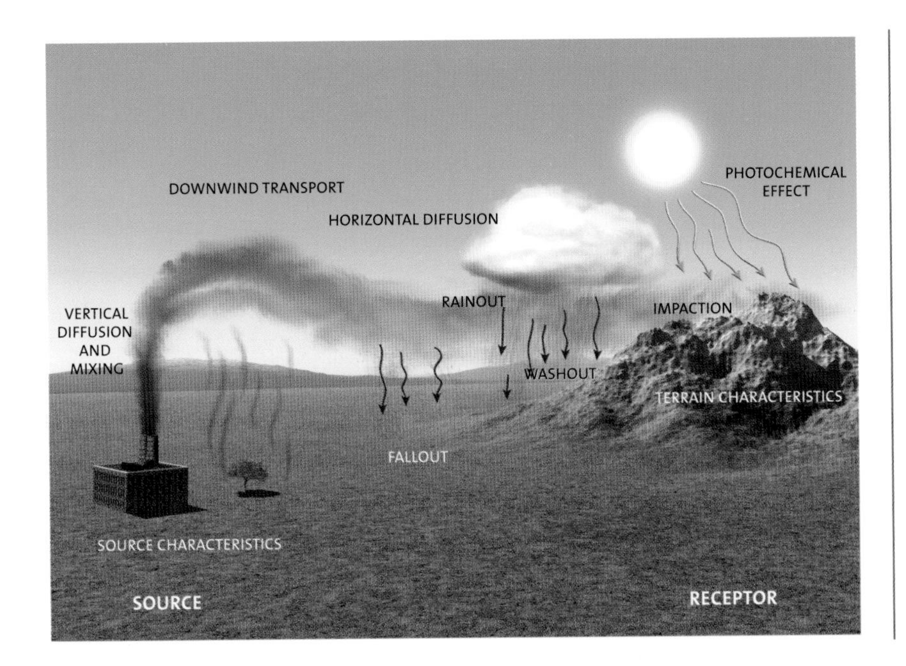

FIGURE 5.14. *Schematic view of the movement of pollution into and through the atmosphere*

with different averaging periods have been established for some of these criteria pollutants. This arises because the atmosphere and the pollution sources can combine to give highly variable concentrations over short periods of time while the human body can have problems with a very high dose for a short period and with a lower dose over an extended time period.

Since we have frequent inversions, there are many occasions when we are in danger of crossing the threshold for one or more criteria pollutant(s) and having air pollution problems. Most of these problems involve a single pollutant from a single source, and in these conditions the impact is usually felt only over an area of a few miles surrounding the source. The scattered and often rural nature of our industrial and energy-generating plants ensures that their pollution generally remains a local problem, even if is significant over that area.

The major exception to this local impact is in the formation of *tropospheric ozone*. This highly corrosive gas is formed when hydrocarbons, commonly those emitted by automobiles, react with various oxides of nitrogen in the atmosphere. Sunlight is required to provide the energy to drive the reaction that produces the ozone. In winter there is not sufficient energy to create problems. In summer, when sunlight is abundant, however, we get the highest concentrations of ozone. At present most of the problems occur in our urban areas, where most automobile trips are made.

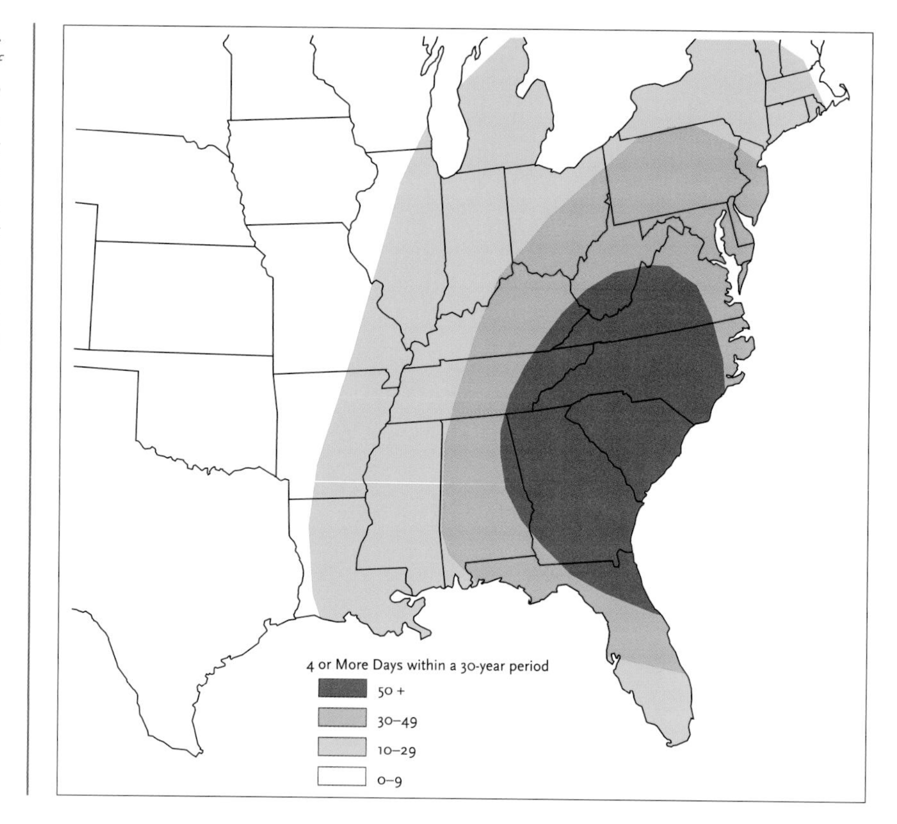

FIGURE 5.15. *Frequency of inversions in the eastern United States (data from J. Korshover, U.S. Public Health Service Publication No. 999-AP-34 [1967])*

But the inversions that drive the meteorological aspect of the process often cover a much broader area. Already some more rural counties are affected. During the worst episodes, as much as half of the state may be affected (see Box 5.5). Since it is unlikely that we humans can significantly modify the frequency of inversions, the problem will continue to increase as our cities expand, unless we take steps to reduce the amount of pollution emitted.

This near-surface ozone, the tropospheric ozone, causes corrosion in almost all materials, including the human body. Clearly such ozone is highly undesirable. However, this same gas high in the atmosphere, the *stratospheric ozone,* is highly desirable. At high altitudes the natural and necessary reactions between ozone and other atmospheric gases are driven by energy from the ultraviolet portion of sunlight. This reduces the amount of ultraviolet radiation reaching the earth's surface. In small amounts this radiation causes sunburn, but as the dosage increases, so does the danger of skin cancers. Thus stratospheric ozone helps keep the amount of ultraviolet radiation that reaches the earth's surface at a level we can absorb without much problem. However, some pollutant gases can penetrate into

TABLE 5.8. Criteria Pollutants

POLLUTANT	MEAN VALUE	PRIMARY STANDARD	SECONDARY STANDARD
Particulate matter—coarse	annual arithmetic	50 $\mu g/m^3$	50 $\mu g/m^3$
	24 hr	150 $\mu g/m^3$	150 $\mu g/m^3$
Particulate matter—fine	annual arithmetic	15 $\mu g/m^3$	none
	24 hr	65 $\mu g/m^3$	none
Sulfur dioxide	annual arithmetic	80 $\mu g/m^3$	none
	24 hr	365 $\mu g/m^3$	none
	3 hr	none	1,300 $\mu g/m^3$
Nitrogen dioxide	annual arithmetic	0.053 ppm	0.053 ppm
Carbon monoxide	8 hr	9 ppm	none
	1 hr	35 ppm	none
Ozone	1 hr	0.12 ppm	0.12 ppm
	8 hr	0.085 ppm	none
Lead	annual arithmetic	1.5 $\mu g/m^3$	1.5 $\mu g/m^3$

Note: $\mu g/m^3$ = micrograms per cubic meter; ppm = parts per million by volume

the stratosphere and interfere with the required reactions, allowing the passage to earth of more ultraviolet light. The gases, especially the *chlorofluorocarbons* developed for use in cooling systems, are emitted worldwide. They are mixed by the winds and distributed over the planet as they drift upward into the stratosphere. The chemistry and meteorology of the stratosphere concentrates their actions into the polar regions, particularly over Antarctica. The result is a marked decrease in ozone there, creating what we have come to know as the *ozone hole* in that region. The size of this hole has been expanding, threatening an increase in ultraviolet radiation and possibly in skin cancers in more heavily populated areas. Less reactive substitutes for the chlorofluorocarbons have now been developed and are slowly being introduced, potentially reducing the impact on the stratospheric ozone and removing the ozone hole.

ACID RAIN AND ITS IMPACT

Acid rain is a regional pollution problem involving pollutants emitted by several states. Gas and small particulate matter can be transported long distances by the atmosphere, and some will be dissolved into cloud water (see Fig. 5.14). The resul-

BOX 5.5. THE OZONE ALERT ON JUNE 25, 2003

As our state develops and adds more industry, more people, and probably most importantly, more automobiles, the chance of severe ground-level ozone pollution increases. Many times each summer we have the right weather conditions for such pollution: lots of sunlight, rather still air, temperature inversions, and high humidity. When these conditions persist for a few days, the hydrocarbons we humans put into the atmosphere can build up and lead to high concentrations of pollutants.

One such event occurred on June 25, 2003. The episode started after a cold front crossed the state in the morning of June 20 and then stalled offshore. Behind the front a small surface high pressure region developed over Lake Michigan. There was a very slack pressure gradient over the mid-Atlantic region, and skies were cloudless. As time progressed, the high pressure center expanded laterally and penetrated deeper into the atmosphere. At the same time, the stationary cold front decayed. On June 23 the Bermuda high joined with the high pressure region to our west, establishing a deep calm region over all of the southeastern United States and much of the adjacent Atlantic Ocean. Cloudless days and nights were the rule, and concern for high ozone concentrations increased. The weather situation persisted virtually unchanged through June 25. On that date the maximum pollution concentrations were recorded, and many places in the Piedmont exceeded the criteria levels. Late on June 26 a front slowly approached from the west, bringing some clouds and considerable vertical motion. Early on June 27 it passed over us, mixing the air, reducing the amount of sunlight penetrating the atmosphere, and lowering the ozone levels. Later that day a rain shower signaled the end of the episode.

While this was not a spectacular event like a hurricane and no deaths could be directly attributed to it, a period of high pollution concentration has a slow, insidious meteorological impact that, if continued, can have widespread health consequences.

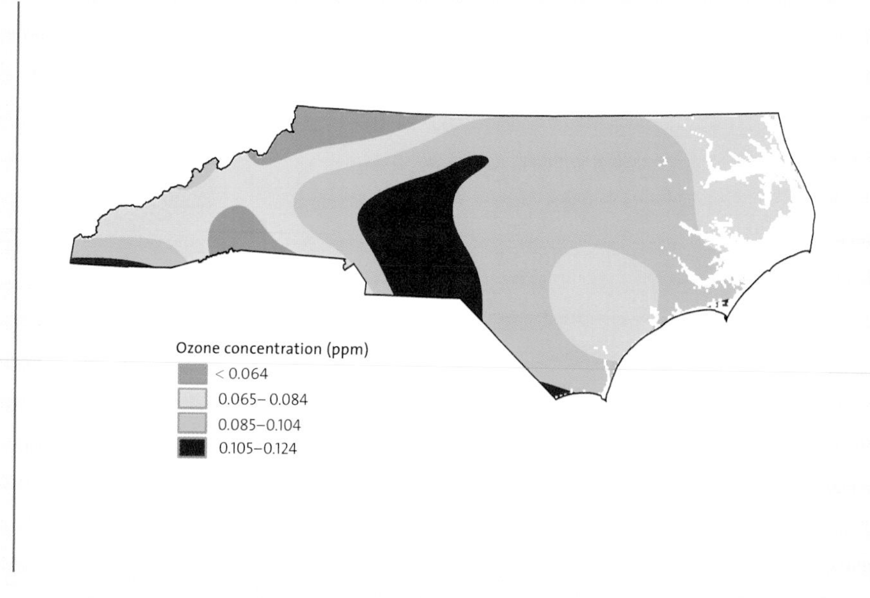

FIGURE 5.B5. *Peak ozone concentrations on June 25, 2003. Much of the urban Piedmont exceeded the national primary standard of 0.085 parts per million over an eight-hour period. (data courtesy of the North Carolina Department of Environment and Natural Resources, Division of Air Quality)*

tant cloud droplets will be more acidic than they would be in a natural, unpolluted atmosphere. Within the cloud the more acidic droplets become concentrated near the base. From here some will return to earth incorporated into raindrops, falling as acid rain anywhere in the state. Some of the acidic raindrops may use an alternate and more localized route and return to the surface by direct deposition when a cloud blows over a mountain range.

For North Carolina at this time the major impact of acid rain occurs through direct deposition. This is limited to our highest elevations, with dead trees being the most obvious indicator. Many of the tops of our high mountain ranges are frequently bathed in clouds. Some cloud droplets are then deposited directly onto the landscape, particularly onto the upstanding trees, as the clouds blow through. This age-old process, which provides a great deal of needed moisture for our high-elevation forests, now ensures that the high-acidity droplets near the cloud base are deposited. This deposition may damage tissue directly or run down stems and trunks and enter the soil. The soil water becomes acidic, and this is, in turn, incorporated into the plants through their roots. Whatever the exact route, it is not a healthy process.

This increased acidity is almost certainly contributing to the decline of the forests in the western mountains. However, there are other factors at work. Many of the high-altitude trees, planted seventy or eighty years ago after the original forest was clear-cut, are old and are sensitive to any kind of stress. Further, various biological agents, especially woolly aphids, are attacking trees and providing additional stress. Finally, the soils and vegetation in the area were developed during the long period of cold represented by the Ice Age. Warming began around 15,000 years ago, and the biosphere is still adapting to the changed, and still changing, conditions. So the cause of death of an inordinate number of trees is not completely clear. Certainly changes in the atmosphere, of both natural and human origin, are involved.

Throughout our state the rain that falls is more acidic than it used to be. In many areas, such as Clinton and Raleigh, the acidity is decreasing (see Fig. 5.16). Rain is slightly acidic even in an unpolluted atmosphere, and for these cities the trend marks a return to natural conditions. This is not the case for other areas, such as Mt. Mitchell, where the acidity is increasing. As with any type of pollution, a trend may be due to changes in the amount of emission or to shifts in the meteorological patterns that move, mix, and deposit the pollution. Most probably for acid rain, it is a combination of both. The decrease in pollution is due at least in part to the effects of air quality regulation and reduced emissions. From a meteorological standpoint, before we can fully analyze changes in patterns we have to answer the question,

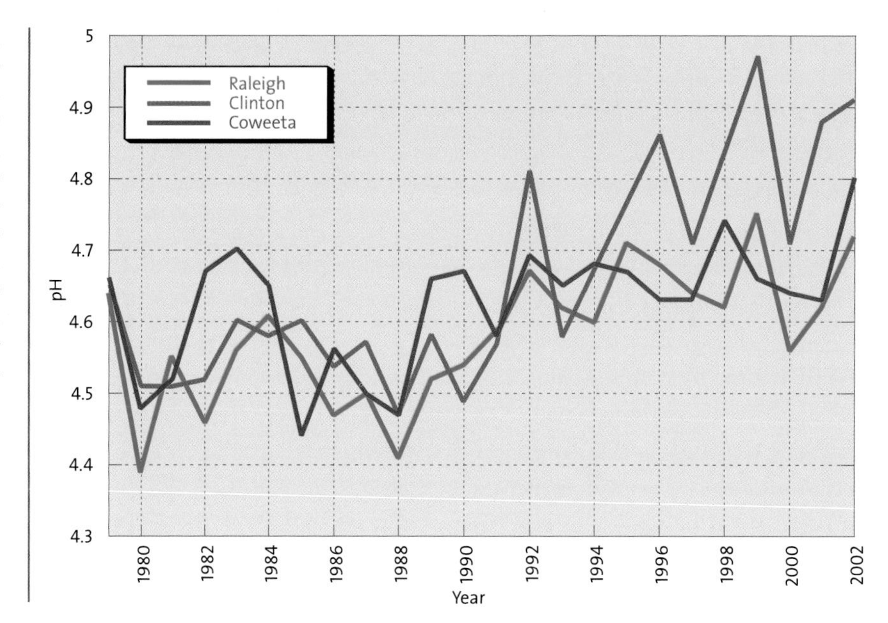

FIGURE 5.16. *Trends in the acidity of precipitation at three sites in North Carolina. The smaller the number, the more acidic the precipitation. Before the industrial era, rain probably had a value close to 6.0.*

"Where does the pollution come from?" The answer is not easy to determine. Certainly it is often possible to stand on a western mountain summit and see a gray-brown layer of particles drifting in from the west. But "backtracking" that air to see what route it took to reach us cannot be done with any certainty. Even if we were fairly precise, we would still have to identify at what point the "gray-brown stuff" was put in. That is even more difficult to determine. We only see one small portion of what could have been a very complex route. That specific small portion may suggest that sources are in the Ohio Valley or involve the industrial plants of Tennessee. These are logical deductions, given the wind directions we commonly experience. But air originating in New England—or even over the eastern portion of our own state and following an almost circular path—might be a source for our pollution on any given day. Local sources will also be involved, and much of the haze, if not the acid rain, may come from nearby power plants and automobile emissions. So while we look to our west and northwest as the immediate source of our pollution problems, in fact what we have is certainly a regional, probably a national, and possibly a global problem.

chapter 6

Weather around the State and through the Year

North Carolina is a state with three geographic regions: the mountains, the Piedmont, and the Coastal Plain. The mountains are relatively cool and wet, the Coastal Plain is wet and warm, and the Piedmont has intermediate temperatures and is our driest region (see Fig. 6.1). Each region has its own distinctive character. In earlier chapters we have considered some regional characteristics, but our concern has been mainly with how they fit into, or differ from, North Carolina's general weather scheme. In this chapter we look directly at regional characteristics.

North Carolina also has four distinct seasons. This may seem obvious to us, but much of the world is not as fortunate and has only wet and dry seasons, or even no discernible difference in weather throughout the year. Part of our good fortune is a product of the diversity of the regions. Fall is the time for color in the mountains. Summer is a time for leisure at the coast. Spring brings fresh blooms to the cities, suburbs, and country of the Piedmont. A clear, crisp cold snap is characteristic of winter everywhere in the state. So here we not only look at the individual regions but also consider the seasonal weather that helps to make them so distinctive.

Mountain Weather: Peak and Valley Contrasts

Probably the major characteristic of the weather and climate of the mountains is the great contrast that can occur in a very short distance. Temperature decreases with altitude, so high points are cold and usually windy. Mountain slopes are often wet, since rain is often created when moist air rises to pass over them. Meanwhile, valleys are often sheltered and calm, and warm and dry. But a change in wind direction can quickly bring a change in the weather. Then wet slopes become dry

FIGURE 6.1. *Monthly (a) average temperature and (b) total precipitation for the three major regions of North Carolina*

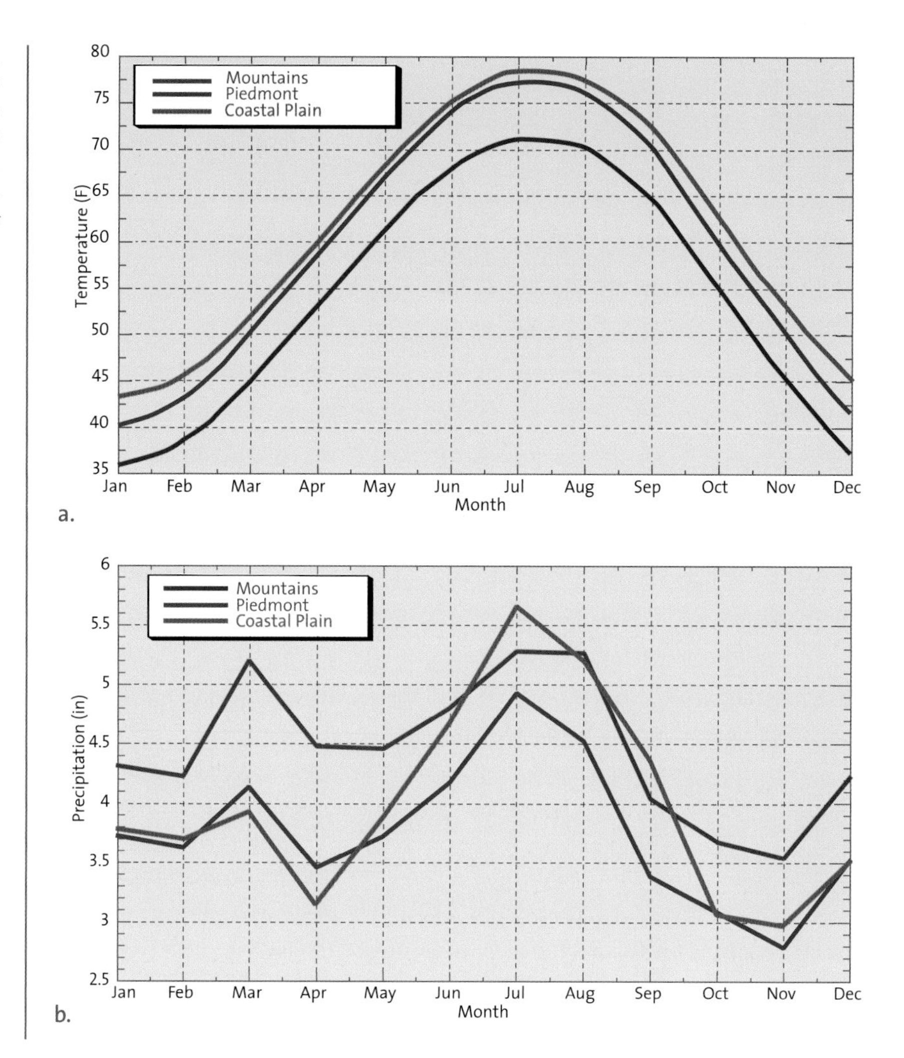

and dry slopes become wet, while temperatures in the valleys may match those on the mountain peaks. So it is difficult to make general statements about the mountain weather and climate.

Another complication arises because we do not have an even distribution of observing stations. Most observations are made by humans where they live, and most people live in valleys, not on mountaintops. While we do have a few mountaintop observing stations, our knowledge of mountain weather is probably more accurately called knowledge of mountain *valley* weather.

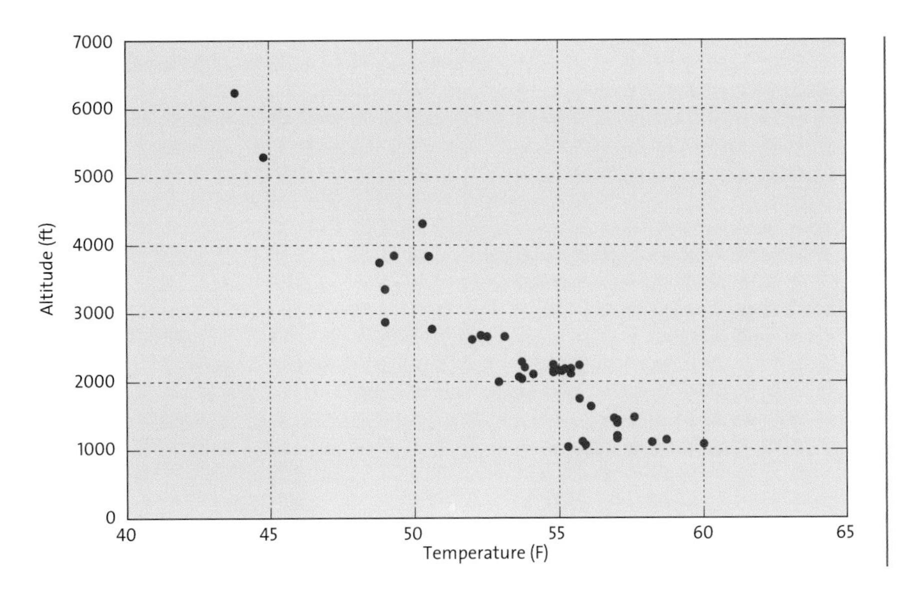

FIGURE 6.2. *Annual average temperatures as a function of altitude for all North Carolina stations having an elevation of more than 1,000 feet*

TEMPERATURE AND ALTITUDE: THE THERMAL BELTS

The temperature observations in the North Carolina mountains clearly show that temperature decreases with altitude (see Fig. 6.2). The exact values at the various stations will vary from day to day, but on most days Mt. Mitchell will vie with Grandfather Mountain to be the coldest spot in the state. The latter currently holds the all-time record (see Box 2.1). But the relationship between temperature and elevation, even on the annual basis shown here, is not simple and straightforward. It was the variations in the observations in this area that led, in the late nineteenth and early twentieth centuries, to a set of practical experiments that increased our understanding of meteorology.

In the mid-nineteenth century Silas McDowell, a mountain farmer, noted that his crops seemed to do well on the valley sides but not in the bottoms or on the ridge crests. He sought advice and information from various local professors. Dr. Joseph LeConte (after whom the mountain in the Great Smokies is named) responded, agreeing that the valley slopes were the warmest places. He called the slopes the "thermal belts" but provided no explanation. However, the name stuck, and now in the area are places such as Thermal City and Isothermal Community College.

The valley-side thermal belts became a favored site for apple orchards, and the region was highly productive and prosperous. Early in the twentieth century the North Carolina Department of Agriculture, in cooperation with the U.S. Department of Agriculture, undertook a measurement program to find out exactly what

FIGURE 6.3. *Processes heading to the formation of katabatic winds*

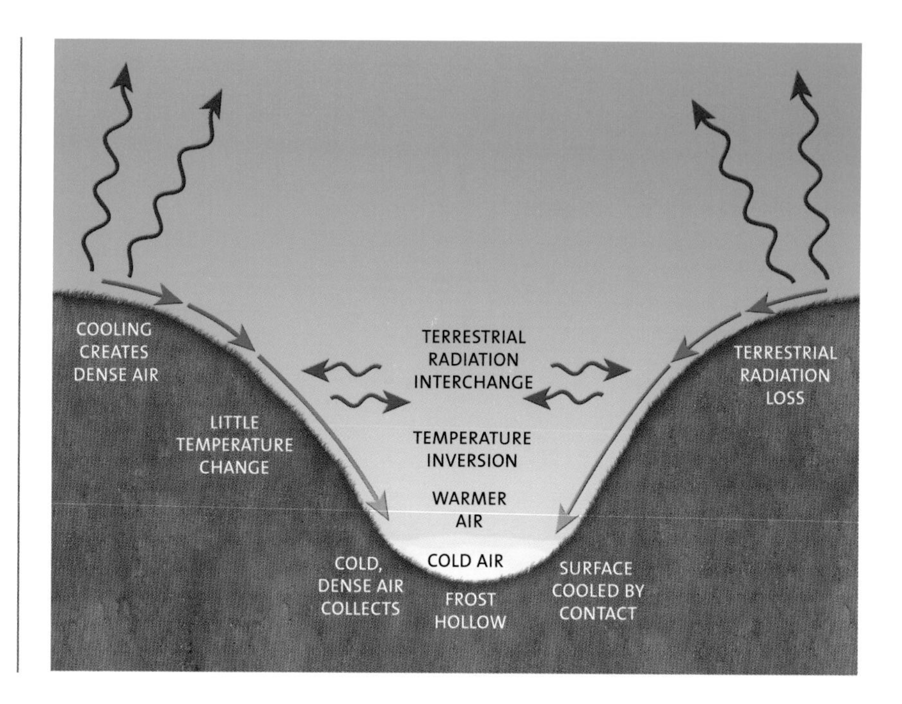

the temperatures were. Sets of observing stations were installed on eighteen hillsides throughout western North Carolina. A unique dataset was developed, and analysis indicated that some, if not all, of the slopes did indeed demonstrate the temperature banding. Later studies in other areas, along with increased understanding of the factors causing temperature changes, identified the processes acting. During the night the emission of terrestrial radiation from the mountaintops created cold summits. The cold, dense air then drained down the slopes as a *katabatic wind*, to collect as a pool of cold air at the base (see Fig. 6.3). Meanwhile, the slopes remained relatively warm. The pool of cold air collecting at the bottom of the valley—most likely to be seen on calm, clear nights—constitutes a frost hollow.

The *frost hollow*, in addition to being a place where frost is likely, is also a place where inversions are common. As the cold air collects in the hollow during the night, it lifts the original warmer surface air, creating the inversion. Any industrial effluent emitted by an industry on the valley floor will be trapped in the inversion. Pollution concentrations will increase as the night progresses, but the daytime warming of the land surface will likely remove the inversion and let the pollution escape.

Frost hollows are by no means restricted to the mountains. Many Piedmont

TABLE 6.1. The Wettest and Driest Places in the State

STATION AND COUNTY	ANNUAL TOTAL (INCHES)
Wettest	
Lake Toxaway (Transylvania)	91.2
Highlands (Macon)	87.6
Rosman (Transylvania)	79.1
Mt. Mitchell (Yancey)	74.5
Driest	
Canton (Haywood)	42.0
Marshall (Madison)	40.3
Enka (Buncombe)	39.8
Asheville (Buncombe)	37.3
Coastal Plain	
Wettest: Southport (Brunswick)	61.0
Driest: Nashville (Nash)	42.9

valleys are cold spots on calm, clear nights. Frost hollows occur even on the Coastal Plain. The valley of the Meherrin River near Murfreesboro has a well-developed frost hollow that provides the right environment for a vegetation community we would normally expect to see in the mountains, not on the Coastal Plain.

WET AND DRY REGIONS

Precipitation also varies widely from place to place in the mountains. This occurs largely because of the orographic effect. Mountains cause all approaching air to rise, which often leads to clouds and orographic precipitation on the upwind side. Clear, dry conditions occur in the lee. On many days, that means that the west-facing slopes get clouds and rain while the eastern ones are dry. But often the situation is reversed, and the east is wet and the west is dry. The only consistently dry places are the bowls in the hills, effectively on the lee slopes for all wind directions. But "dry" is a relative term, since much of the rain everywhere in the state comes from cyclonic storms, which are not directly affected by topography. So stations in the broad basin surrounding Asheville may get 40" annually. This is dry compared with stations in the nearby hills, which may get 80" in a year (see Table 6.1), but it does not suggest desert conditions.

MOUNTAIN WEATHER THROUGH THE YEAR

The temperature cycle through the year is virtually identical throughout the mountains (see Fig. 6.1a). Altitude is the major factor that determines the actual values each month. The precipitation pattern, in contrast, shows no such regularity from station to station. The local topographic conditions, interacting in different ways with the airflows bringing the rain, have an overriding influence. For specific places, the information in Appendix C should be used. One of the few generalizations possible is that almost all stations have at least three inches of rain in all months, and there is no dry season.

Winter

Temperatures are, not surprisingly, low in winter, and freezing conditions are more common in the mountain region than elsewhere in the state. Snow is also more common (see Fig. 3.14). This may lead to more problems with hazardous travel and school closings, but it also creates the opportunity for a ski industry. However, only a small portion of the precipitation falls as snow. Much of the mountain precipitation is relatively warm rain that tends to melt any snow. So the ground, even at high elevations, is rarely completely covered with snow for long periods during the ski season. Resorts have to make artificial snow. The snow-making guns rely on cooling by expansion to turn ejected water droplets into falling snowflakes. Fortunately, the mountain air is usually sufficiently cool and moist to allow this, and artificial snow can be made on many days. In most years, therefore, a continuous snow cover can be maintained throughout the heart of the season. In mild years, or years with more than average rainfall, there are likely to be extended periods when the snow is melting and conditions are not suitable for making snow. Our mountain climate seems to be just a little too warm to guarantee good skiing throughout the season every year.

Spring

Anyone driving into the mountains from the Piedmont in early spring cannot fail to notice that the buds burst into flower later the higher you go. This phenomenon depends on many factors other than weather—plant species, soil type, and slope exposure, for example—but climate plays a great role. We can summarize the effect by looking at the accumulation of growing degree days early in the year at a line of stations up the Blue Ridge roughly along the route of U.S. 421 (see Fig. 6.4). The growing degree day, the number of degrees Fahrenheit the average temperature is above 40°F each day, is a rough measure of energy available for plant

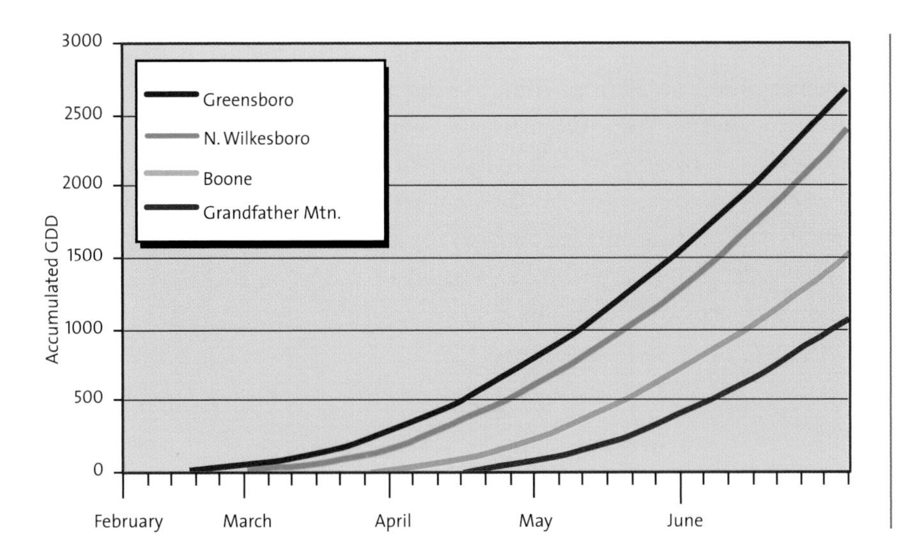

FIGURE 6.4. *Average accumulated growing degree days as the growing season progresses at four stations along a transect up the Blue Ridge from Greeensboro to Grandfather Mountain. At the lower elevations the growing season starts earlier and there are more growing degree days.*

growth, so a rapid accumulation suggests early development of the plants. There is relatively little difference between Greensboro at 897 feet and North Wilkesboro at 1,120 feet above sea level, respectively. Five hundred growing degree days are accumulated on April 12 and 21, respectively. But Boone, at 3,360 feet, needs until May 16 to accumulate that amount. Grandfather Mountain, at 5,300 feet, does not achieve that total until June 2. So virtually all aspects of plant development are delayed.

Summer

Since colonial times it has been the ambition, for those who could afford it, to get away from the malarial and "unhealthful" summer air of the Coastal Plain—and even much of the Piedmont—and into the mountains. Charlestonians, for example, moved into the Tryon (Polk County) region at the base of the Blue Ridge during the summers of the early nineteenth century. Better transportation soon allowed them and others to spend the summers atop the ridge, with Flat Rock and surrounding areas providing the setting for new homes and resorts. Soon after the Civil War the whole area, Asheville in particular, developed a tourist industry based on the favorable summer weather. Numerous publications, some of which included essays by local physicians, encouraged visits for rest and recuperation in this healthful climate.

The mountain air is certainly cooler than the air of the lowlands. Figure 6.1a suggests that the difference is commonly several degrees Fahrenheit. There is also less moisture in the atmosphere (see Fig. 6.5); Asheville has lower average dew

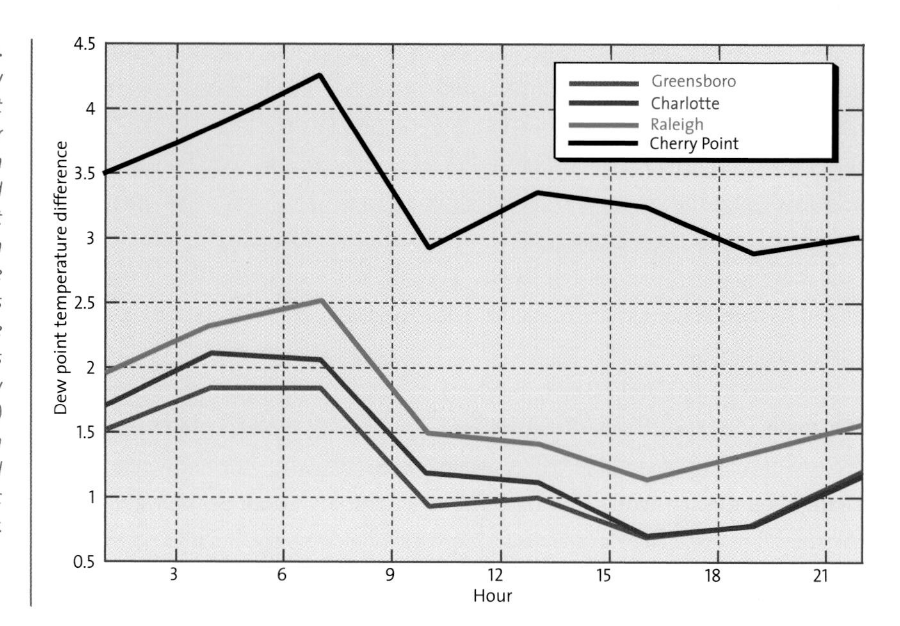

FIGURE 6.5. *Difference in July average dew point at three-hour intervals between Asheville and selected Piedmont and Coastal Plain stations. The Piedmont is somewhat more humid; all stations there have dew points 1–2°C (2–5°F) higher than those in Asheville. The coastal dew points are 3–4°C (5–7°F) higher.*

points than any other observing station in the state. This difference is particularly marked between Asheville and Cherry Point (Marine Corps Air Station, Carteret County) on the Coastal Plain. For all stations, however, the differences are greatest in the morning. It is possibly this dry morning air, combined with the overall lower temperatures and lower humidity, that gives the more healthful, invigorating feel to the summer mountain climate. Certainly many other aspects of the weather, such as sunshine, clouds, and wind, do not differ greatly between the various regions of the state. Nonclimatic factors, of course, including psychological and aesthetic ones, probably play a major role in the perception of the mountains as a healthful vacation region.

Fall

Although the leaves change color in the fall throughout the state, only in the mountains do the spectacular reds, browns, and golds that provide a tourist attraction occur. In part, this is because of the presence of a richer range of deciduous tree species in the area, in part because of the vistas afforded by the mountains themselves, and in part because of the rapid onset of fall and the probability of early frosts.

Virtually all deciduous trees respond to the decrease in day length after the Summer Solstice by starting chemical changes that lead to changes in leaf color. Some trees, such as birches, tulip poplars, redbuds, and hickories, turn yellow whatever the circumstances. Others, such as sugar maples, dogwoods, sweet gums, black gums, and sourwoods, usually turn red. But in order to do that, they need plenty of solar radiation—sunshine—to stimulate the leaf pigment that gives them the red color. Otherwise they will turn yellow. In addition, the temperature, and especially the occurrence of frost, helps determine when the changes will take place. One great characteristic of our southern Appalachian mountains is that there is a tremendous mix of tree species, so that—unlike New England, for example—we have a great range of colors and an extended period when those colors are present.

The time of colorful trees commonly stretches from late September to early November, but it varies from year to year. Forecasting the best time to see the fall colors very far in advance is a risky business. Small temperature fluctuations can change the timing, while slight variations in cloud amount at the right point in the tree's life cycle can have a major impact on just how spectacular the display is to be. In 2001 the peak color occurred around the middle of October in the north and slightly later in the western mountains and on the higher peaks. The peak occurred about two weeks later in 2002, mainly as a result of the delay in cooling in the fall. In addition, the higher solar radiation amounts in the latter year provided some better colors—as well as more sunny days for visitors to enjoy the view.

The Piedmont: Town and Country Contrasts

The Piedmont is a region of rolling hills with frost hollows and orographic effects akin to those of the mountains. The effects, however, are much less marked. At the same time, it is also an agricultural area with a mixture of cropland and forest, and wet areas and dry patches. However, these landscape differences are less clearly defined than they are in the Coastal Plain. Probably what makes the Piedmont meteorologically distinct is the presence of the major cities of the state. So our emphasis for the Piedmont region weather is on the interaction of humans and their structures with weather and climate.

THE CLIMATE OF CITIES

The building of a city changes virtually all aspects of the weather. The most characteristic change is the development of an "urban heat island," with temperatures within the city being a few degrees warmer than the surrounding rural areas. The detection of this effect requires special observation programs. For North Carolina

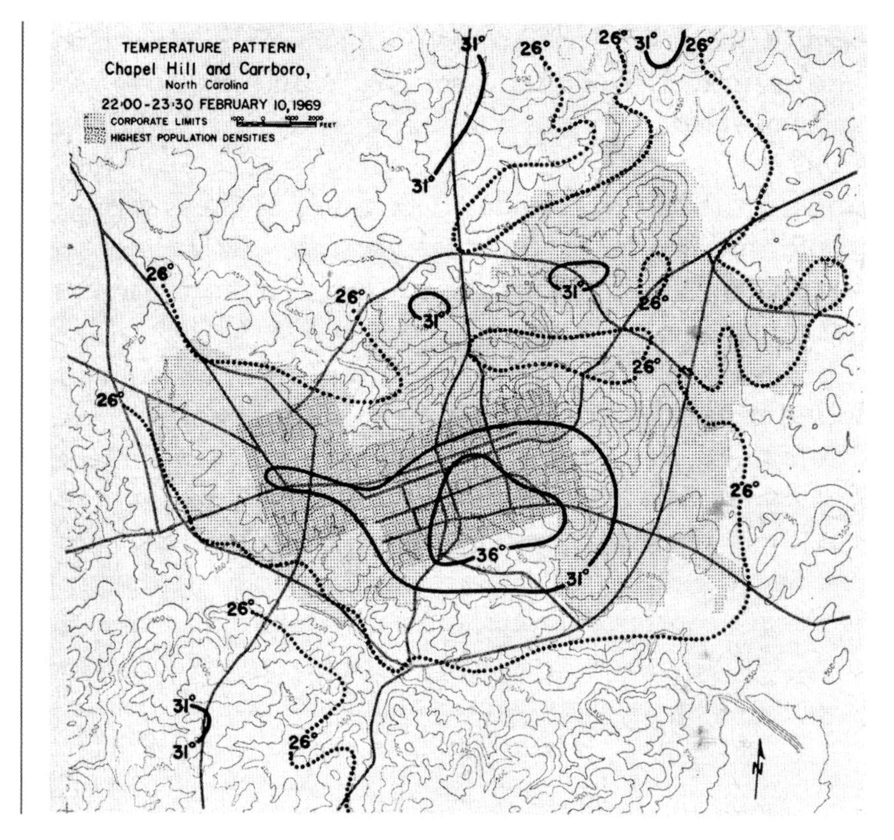

FIGURE 6.6. *The Chapel Hill urban heat island recorded on a winter evening in 1969 (reproduced from the* Bulletin of the American Meteorological Society *[1970], 604, courtesy of the society)*

this has been explored in detail only in Chapel Hill (see Fig. 6.6). This exploration involved a series of automobile trips across the city. A recording thermometer, shielded to make sure that engine heat did not affect the sensor, was attached to the front of the car. The temperature was recorded as the car proceeded across the city. Chapel Hill—especially in 1969 when these observations were made—is a small city with lots of grass and trees even in the downtown area. The effect was not as marked as in some other cities, such as Chicago or London, but the downtown core on the night shown here was about 10°F warmer than the area outside town.

Although it is difficult for us to detect the urban heat island effect without special instrumentation, we often see the result. Many plant species in many years blossom in the central parts of cities a week or so before they do in the suburbs. And the suburbs themselves are in advance of the rural areas.

It is not really surprising that cities change virtually all aspects of weather and climate. City air is different, usually less clean and clear, from rural air, so pollution levels are higher. Surface types also differ. The replacement of rural soil and vegetation, which can cool by evapotranspiration, with dry urban concrete that has no opportunity to cool by evaporation leads directly to the urban heat island. The hard urban surfaces also encourage the development of more frequent floods.

Wind flow is also influenced because the mix of buildings, roads, and open spaces in a city creates a rougher surface than the forest and cropland mix of rural areas. This means there is more friction on the wind, and the average wind speed in the city is lower than that in the countryside. But variability from spot to spot is increased. The buildings themselves, especially those downtown, act as obstacles to the wind. For some streets this leads to the funneling of the wind and a marked increase in wind speed; Tryon Street in Charlotte or Fayetteville Street in Raleigh can be very windy indeed. Cross streets may have almost calm conditions, often accompanied by the smell of increased pollution concentrations. A slight change in wind direction may lead to a reversal of these conditions.

These weather changes related to cities are currently small in effect and local in scale. Even the largest of our North Carolina cities have only small areas where trees and grass are completely absent, and anyone flying over our state, even the densely populated Piedmont, gets an impression of a heavily wooded or cultivated landscape with a few isolated urban centers. The atmosphere, as it blows over, "sees" the same thing. So as yet, we have not significantly changed the surface of our state as far as the weather is concerned. This also seems to be the case for even the largest cities of our planet. They change the local weather but do not have an impact far outside the confines of the city. But we continue to change the nature of the surface for ever larger areas, and at present we are not sure how big a change is needed to produce global, rather than local, weather and climate effects.

MODIFYING LOCAL WEATHER FOR OUR BENEFIT

By building cities we have inadvertently modified our weather and climate. Deliberately modifying climate for our benefit has proved more difficult, and only in a few isolated instances have our efforts been reliable and effective. The building of houses for shelter and comfort and the installation of irrigation systems to modify the rainfall regime have been so successful that we rarely think of them as weather modification techniques. They have been adopted everywhere in our state. Two other, more specialized modification techniques have also been adopted: frost protection for agricultural crops and *shelter belts* to reduce wind speed for various reasons.

Over the centuries, housing styles in an area have developed to reflect the building materials and technologies available, while the design has been, in part, a response to the weather. Houses in snowy climates tend to have steeply pitched roofs to shed the snow, and desert dwellings usually have thick walls with only small openings, shutting out the heat of the day and the cold of the night. Our climate needs houses that protect us from the heat and humidity of summer, from the cold of winter, and from rain at every season.

Shade in summer can be provided by a large roof overhang. While preventing the penetration of sunlight into the house in summer, it allows the rays of the low winter sun to enter, just when we need them for heating. The roof itself, of course, is vital for keeping the rain out. What we do want inside is a breeze to stir the summer air and push the heat out. So lots of openings are helpful, particularly if they allow a free flow of air through the whole house. But we do need to be able to close them off in the winter to prevent the cold winds from entering. Put all this together in a practical fashion and what do we get? The white-painted, steep-roofed house with a porch on all sides—the house regarded as a typical symbol of the South. This design, particularly when the front and back doors are aligned to maximize the through breeze and when deciduous trees are located nearby to provide extra shade in summer, represents the most effective method of weather modification for our comfort in our climate. The wraparound porch is almost a free bonus, providing extra space to catch summer breezes while giving extra protection from colder winds in winter. The swing on the porch is not obligatory, but it adds to the cooling effects of the breeze.

This traditional design minimizes the heating and cooling energy required for a house. Even in colonial days some winter heating was required, but air conditioning was unknown. Summer was therefore not too pleasant, and people with sufficient resources moved to the cooler mountains. But the houses were bearable as long as the pace of life was not too hectic. And the energy bill was zero. The more recent widespread adoption of air conditioning and the abundance of cheap energy means that it is possible to survive the summer in a vastly expanded range of housing styles. Our climate has not changed significantly since colonial days, so the farther the design of a building is from the traditional climate-friendly one, the greater the energy needed to maintain it as a comfortable dwelling.

The other common way in which we deliberately alter the weather for our benefit is through irrigation, as detailed in Chapter 3. Although attempts at cloud seeding, also mentioned in Chapter 3, are continuing in various parts of the world, for us irrigation is more reliable and more precise as a rainmaking process. Most of our irrigation systems are used to provide precise amounts of water at a prescribed time to a broad area of cropland. For smaller areas with low-growing and high-value crops, drip irrigation to individual plants, often combined with the use of black plastic ground covers to minimize evaporative losses, is becoming increasingly common.

Irrigation is also used as one of two methods for frost protection. Irrigation can be employed when any type of frost is forecast, but it works only on crops that can be completely coated with a water film and can tolerate a small amount of frost. Strawberries and similar low-growing crops are good examples. When the water freezes on the plant, the change in state from liquid to solid releases latent heat. Much of this will go into the enclosed fruit. The resulting ice coating is a good thermal insulator and prevents heat loss from the warm fruit to the cold exterior. The fruit, in contact with ice, will cool until it is just below freezing and remain that way regardless of however much frost is occurring outside. A problem arises only if there is not a complete coating around the fruit; then energy will flow out through the gap, and the fruit will be damaged.

For crops taller than strawberries, a second frost protection method is available, but this can be used only for radiation frosts. Radiation frosts occur on calm, clear nights early or late in the growing season. The process is similar to that for radiation fog formation considered in Chapter 3. But the air is dry, and so the near-surface temperatures can fall below freezing without condensation occurring. A temperature inversion is created, with warm air lying above the freezing air. Frost can be prevented by mixing the air and equalizing the temperature. Fans—usually airplane-type propellers—do the job. When the method was first invented, there was controversy over whether the blades should be aligned horizontally or vertically. It does not matter; both work equally well.

A final deliberate weather modification is the construction of shelter belts to decrease wind speed. While a solid wall provides some shelter immediately downwind, the air that is forced up and over the wall comes back to earth a little way downstream as a very turbulent flow, making conditions worse. A permeable barrier is more effective. A line of trees acting as a barrier is common. The trunks act as obstacles that let some wind through. This wind is slowed because of the friction. The rest of the air is forced over the barrier, but it cannot return to earth because the air that has flowed through acts as a cushion. Hence there is an overall slowing. Shelter belts of trees are usually constructed to provide protection for houses in open areas. In North Carolina we recognize them most readily around houses on the open spaces of the outer Coastal Plain, where they may be the only trees on the landscape. Smaller shelter belts made of vertical wooden slats are also used commonly on the coast itself to prevent sand dunes from moving and covering our roads.

A PIEDMONT METEOROLOGICAL YEAR

The Piedmont usually has daily average summer temperatures in the high 70s and daily winter temperatures in the high 30s (see Fig. 6.1). There is little difference across the region. Most differences occur on a day-to-day basis as fronts cross the area bringing relatively low temperatures on one side and warmer temperatures on the other. Precipitation in all months of the year is commonly around four inches. In most regions spring and summer are the wettest seasons, and although fall and winter may be drier, they are still rather wet.

Winter

Winter is the time of prolonged, gentle, and usually cold rain. Snow is possible, but it is nowhere near as frequent as in the mountains. Amounts are usually light, and snowstorms tend to be inconveniences rather than dangerous events. But ice storms are a different matter (see Box 3.1). All of North Carolina gets freezing rain events from time to time (see Fig. 6.7). But they are most frequent in the Piedmont, especially in an area extending from southwest to northeast just east of the mountains. This location is no coincidence. The mountains act as a dam, causing cold near-surface air blowing in from the east to collect in an ever deepening reservoir. Liquid raindrops falling from the clouds above are cooled as they pass through this layer. Eventually the layer becomes so deep and cold that the drops are cooled below the freezing point. The major characteristic of this supercooled water is that it freezes as soon as it hits any object on the surface. Freezing rain results.

Even with our modern observational system it is very difficult to measure in fine enough detail the temperature and precipitation patterns involved in an evolving ice storm. So forecasts are likely to be imprecise. However, our understanding and forecasting ability regarding this meteorological phenomenon are increasing as we combine atmospheric theory and computer power in the development of mesoscale models. Models in general are computer-based calculations of the future weather using the sets of mathematical equations that express the processes controlling the weather, starting from observations of the current weather. Mesoscale models commonly deal with a particular feature or forecasting problem on the regional scale. Ice storm forecasting is one example; several others will be mentioned later.

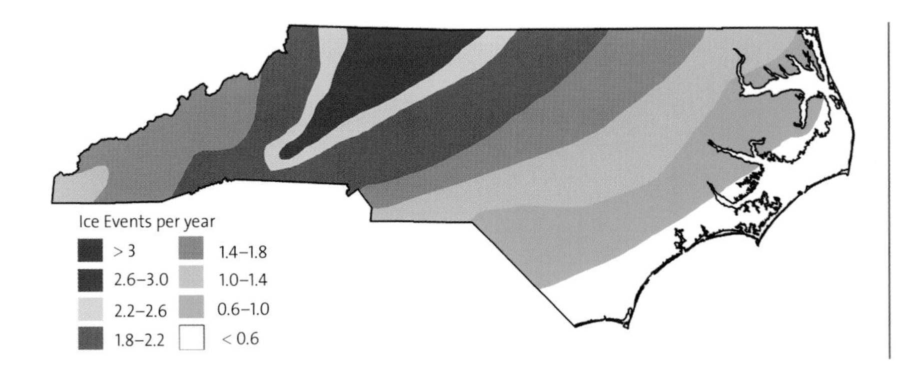

FIGURE 6.7. *Number of freezing rain events per year. The data are based on a model that uses the actual observations from the First Order weather stations and estimates values for the Cooperative stations. The observations at Greensboro show the highest incidence, and this, along with other stations in other states, indicates the major ice storm belt in the western Piedmont. (data courtesy of C. E. Konrad, 2005)*

Spring

Spring in the Piedmont brings a burst of growth from much of the vegetation (see Fig. 6.8). It also brings pollen. Pollen production is a natural part of the annual cycle for most vegetation. Thus some species or another is producing pollen at virtually any time between early spring and late fall (see Fig. 6.9). But spring is the season when most tree species, including some that are prominent on the Piedmont landscape, are most active. Oak is a prolific producer, and some pine species can be almost as troublesome from a human perspective. As the year progresses, the impact of tree pollen declines, but grass becomes a major source, to be replaced, in turn, by weeds as fall approaches. Not until late October for most of our state are we free from pollen.

In most cases the amount of pollen in the air depends on a couple of atmospheric factors. First is the overall nature of the early spring weather, which influences the duration of the pollen-producing period. Commonly a cool, wet spring delays the onset of the pollen season but leads to a shorter, more intense period during which all the trees seem to produce at the same time. Second, the synoptic situation controls the day-to-day concentration of pollen in the air. Windy conditions encourage trees to release the material, but calm conditions with inversions maintain the highest near-surface concentrations. Therefore almost any day in the pollen season is going to present a problem for a sensitive individual.

Summer

Summer is the time for outdoor activities. Often it seems that during the workweek the weather is wonderful, bright, and sunny. Come the weekend, the rain sets in. But—fortunately or unfortunately—the data do not support that assump-

FIGURE 6.8. *Spring blooms in the sunlight of one of the last periods of calm, clear, continental polar air conditions as winter retreats.*

tion (see Fig. 6.10). There is very little correlation between the day of the week and the probability of precipitation. Fall, not surprisingly since it is the driest season, has the lowest probability of rain on any day of the week for most of our state. But there are no other reliable patterns. In summer in Burlington, it rains about one day in three, regardless of what day of the week it is. But in fall, Saturdays seem considerably wetter than Sundays. Other places seem to have different patterns, but there is no clear trend statewide.

Fall

Although climatologically fall is the transition between summer and winter, the transition is rarely smooth. Cold spells early in the season give way to warm periods that, in turn, are replaced by newer, colder spells. Human experience, long tradition, and folk wisdom have created the notion of an *Indian summer*, a period

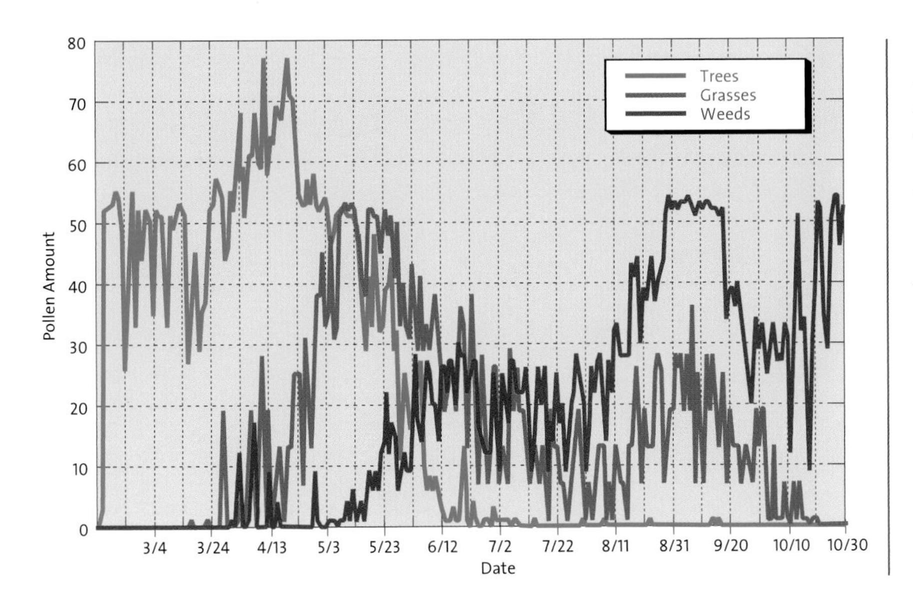

FIGURE 6.9. *Pollen amounts as a function of time of year in the Winston-Salem area. Results, based on several years of observation, are grouped into three major classes, and the amounts are expressed on a relative scale. Higher numbers indicate that more people are likely to be affected, and these may be affected more intensively. (data and scale courtesy of Forsyth County Environmental Affairs Department, Winston-Salem, N.C.)*

of abnormally warm weather, usually with sunny skies, in the middle of fall. This concept appears throughout much of Europe and the eastern United States. In New England it dates back at least to the time of the American Revolution and refers to a warm spell that occurs after the first killing frost of the season. That definition is rather vague and certainly is not applicable to Piedmont North Carolina, where frost may not occur until the end of the year. Nevertheless, we do get Indian summers in many years, so we have defined it as a period of at least three days with temperatures over 80°F immediately after a cool period of ten or more days with the maximum below 80°F, and including at least two days when the temperature never reached 65°F. Using this definition, and the century-long temperature record at Mt. Airy, we find that twenty-eight years had Indian summers. These were clustered in mid-October, but a couple occurred late in September, several came very late in October, and a few were in early November. So it does seem that we get a clearly marked Indian summer about one year in four. In this instance, folk wisdom is supported by our weather observations.

The Coastal Plain: Land and Water Contrasts

Local differences in the weather across the Coastal Plain are not as easy to detect as those of the mountains, and unlike the Piedmont there are no major cities to

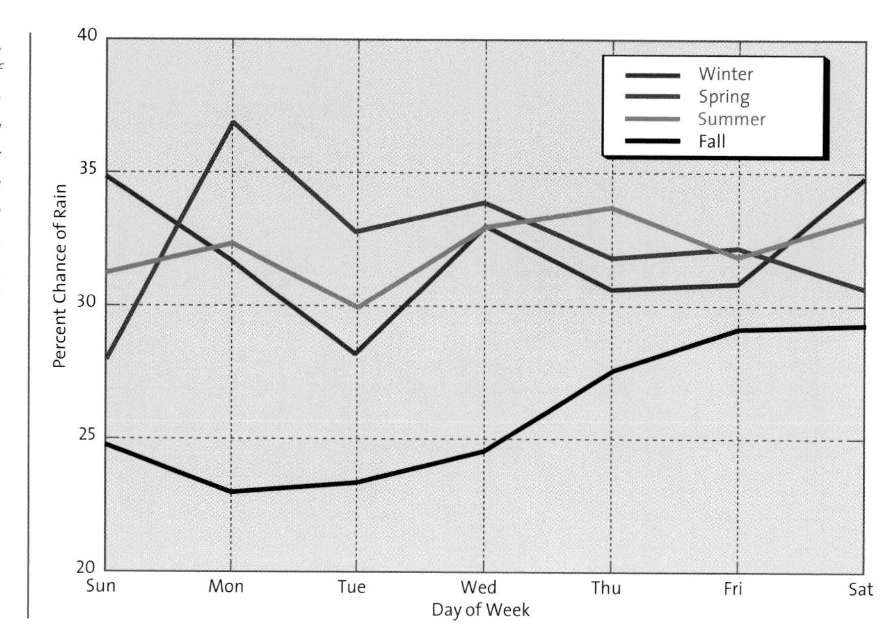

FIGURE 6.10. *Probability of precipitation in Burlington by day of the week for each season. While there seems to be some pattern, especially in winter, no explanation is readily apparent.*

create their own climate. But there are some major differences, mainly a result of the mix of land and water in the region. The land-sea contrast is the most obvious and leads to the development of sea breezes. But on the land itself, the mix leads to local variations in the surface energy and water budgets. These variations may influence the weather and potentially play a role in tornado formation and hurricane modification. So the results of seemingly small differences may be much more spectacular than in the mountains.

SEA AND LAND BREEZES

The contrast between land and ocean surfaces is sufficient for the creation of local coastal air circulations called land breezes and sea breezes. *Sea breezes* are a daytime phenomenon generated because land and sea respond differently to the daily input of solar radiation. Early on a cloudless morning in calm or near-calm conditions, the coastal land will warm rapidly in response to the increasing solar intensity. The nearby water, in contrast, will warm very slowly. As the day progresses, the temperature of the land and the depth of the warmed column of air above it will become much greater than the corresponding temperature and air column over the water. This will set up a pressure difference, with higher pressure over the water and lower pressure over the land (see Fig. 6.11). Eventually air will start

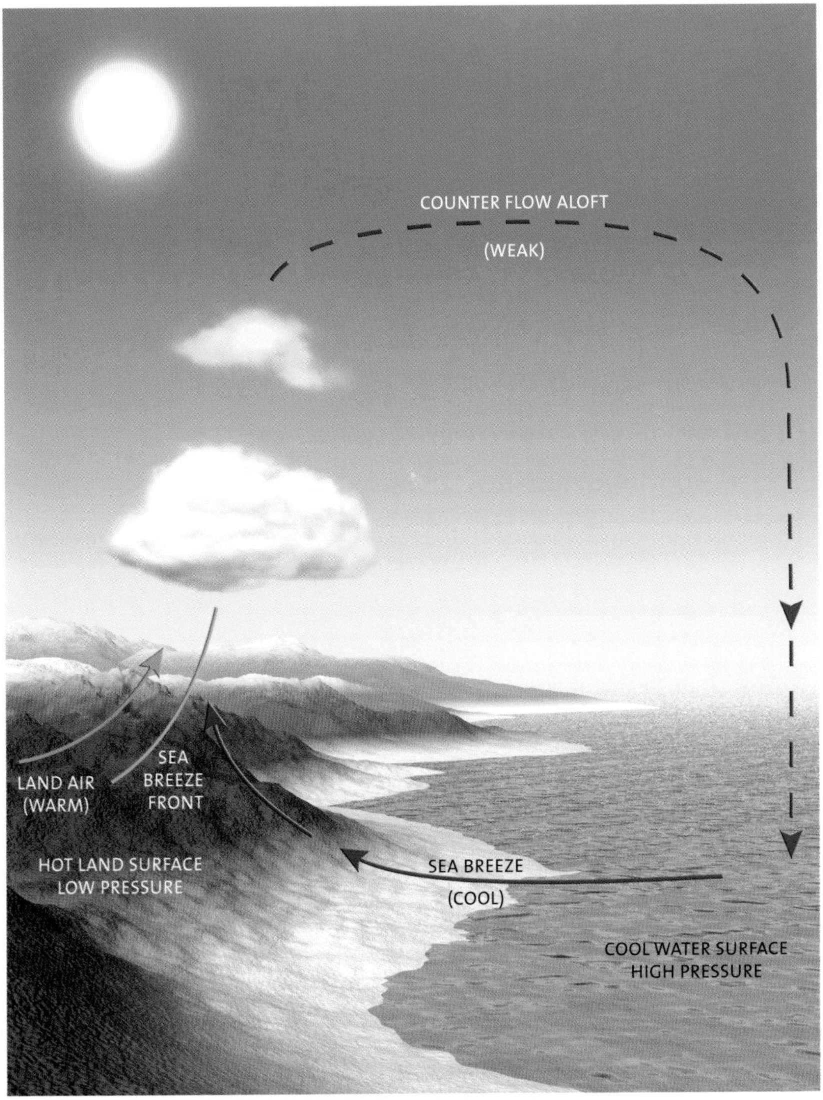

FIGURE 6.11. *Schematic diagram of the formation of sea breezes*

to move toward the lower pressure. This is the sea breeze, a cool, gentle airstream coming off the water. Ahead of it is the sea breeze front, the line of temperature change where the cooler air undercuts the warmer land-based air. The warm air may already be unstable, so the front may trigger convection and cumulus clouds may form (see Fig. 6.12). Commonly in North Carolina these clouds will drift inland, sometimes for forty or fifty miles as the afternoon progresses. They will die

FIGURE 6.12. *Cumulus clouds associated with the sea breeze front some ten miles inland of the Brunswick County coastline. Many of the clouds had drifted overland in response to a light breeze above the surface. They then re-evaporated into the warm atmosphere.*

away in the early evening as the solar energy driving the land-sea temperature difference decreases.

Sea breezes are primarily a summer phenomenon because of the energy needed to create them. They are also generally restricted to our coastline between Morehead City and the South Carolina border. North of this the complex intermixture of land and water, barrier island, sound, field, and swamp offers little opportunity for formation of the required temperature contrasts. But in the south the sea breeze has a significant cooling effect when it passes over the coastal counties, while the sea breeze front can foster rapid uplift that may produce rain or even, very infrequently, thunderstorms and severe weather. The vigor of the uplift on any day depends not only on the land-sea temperature contrasts but also on the interaction of the sea breeze system with the broad-scale weather affecting the entire southeastern United States. This is another complex forecasting problem that the use of mesoscale models is helping to solve.

In theory, the *land breeze* is the nighttime counterpart of the sea breeze, with a cold land and a warm ocean. But in our area land-sea temperature contrasts at night are usually small, since high humidity tends to discourage much terrestrial radiation loss from surfaces of any type. Without the temperature contrasts, there is no circulation created. So land breezes rarely form.

SOIL, VEGETATION, AND WEATHER

Most surface contrasts on the Coastal Plain are not as marked as the land-sea differences. The Coastal Plain is an area with an intermixture of field and forest, of swampy land and well-drained soils. Most of it is well watered, but some areas are much drier. This is particularly true of the Sandhills region, which provides a contrast clearly linking soil, vegetation, and weather.

The Sandhills region of North Carolina, stretching from Fayetteville and Fort Bragg west past Southern Pines and Pinehurst and south to the South Carolina border, is unique within our state because the soil is predominantly sandy. The region lies between the Piedmont and the Coastal Plain, where almost all soils have a much higher clay content. Sandy soil drains very quickly, and even a few minutes after rain, it appears dry. A soil with more clay dries much more slowly and often allows puddles to remain long after the rain has passed. Vegetation in the Sandhills has adapted to this alternate wet/dry soil, and pine forests predominate. The other areas have mixed pine and hardwood forests and a much higher proportion of agricultural land.

The dry Sandhills region acts rather like a city surface, but without the complications such as pollution, building heating, or wind channeling that a city produces. In particular, there is a kind of Sandhills "heat island." Daytime temperatures, especially on sunny days, tend to be higher in the Sandhills than in the surrounding areas. It is not happenstance that, as noted in Chapter 2, Fayetteville holds the high-temperature record for the state. But the surface type also leads to more rapid cooling during the night, and the area tends to become colder at night than the surrounding areas. When averaged over the day, temperatures tend to differ rather little from the surroundings.

The Sandhills region also has the reputation of being North Carolina's tornado alley. The reason usually given is that the air above the hot Sandhills surface is more unstable than that elsewhere, and thus it fosters the development of thunderstorms that can develop tornadoes. This is plausible, but thunderstorms and tornadoes occur in many other areas (see Fig. 5.9). Individual patches of wet and dry soils within the Coastal Plain may produce local hot spots, but the patches seem to be too small to produce thunderstorms. Some recent investigations using mesoscale models, however, suggest that the patchiness leads to variations in wind speed and direction near the surface. In some cases two airflows may converge over a hot spot and encourage the vertical motions needed to start the thunderstorm. The same exploratory mesoscale modeling techniques are also being used to investigate links between differences in surface type on the Coastal Plain and how hurricanes behave as they pass over land.

Thus meteorological interest in our Coastal Plain and adjacent waters is increasing rapidly. Not long ago the area, apart from some interest in sea breezes or tornadoes, was regarded as one where little could be said about the local meteorology. Now it appears that we are beginning to see a lot of connections between the surface and the atmosphere in the Coastal Plain. Time will tell whether that renewed interest can lead to improved storm forecasts for the area.

MARCH OF THE SEASONS IN THE COASTAL PLAIN

The presence of the nearby ocean plays a significant role in the seasonal climate of the Coastal Plain. Proximity to the ocean, where temperatures change little with the seasons, usually creates relatively warm winters and cool summers. However, in our case, this effect is modified because our airflow is often from the west and brings continental air, which is cold in winter and hot in summer, over us. As a result, only along the coastline immediately adjacent to the ocean are seasonal contrasts dampened. For the rest, the seasons are very similar to those of the rest of the state (see Fig. 6.1).

Winter

Climatological data suggest that most of the Coastal Plain gets only a very small amount of snow in winter. The area averages less than 2" along the southeast coastline, but the amount increases northwestward to more than 6" (see Fig. 6.13a). For comparison, the highest mountain peaks get more than 40". But snowfall averages can be misleading. Most winters have little or no measurable snow; others may have one or several major disruptive storms. Along the coast itself there is less than a 5 percent (one in twenty) chance that a storm leaving 8" or more will occur in any particular winter (see Fig. 6.13b). At Cape Hatteras the record suggests that the likelihood is closer to one chance in a hundred. Farther inland the probability increases until it is about 10 to 15 percent at the Piedmont boundary. The probability does not increase very much farther inland across the Piedmont. The Coastal Plain may get less snow than the Piedmont on average, but when big snowstorms arrive, they often affect both regions simultaneously. Some, like the Christmas storm of 1989 (see Box 1.1), affect the coast alone.

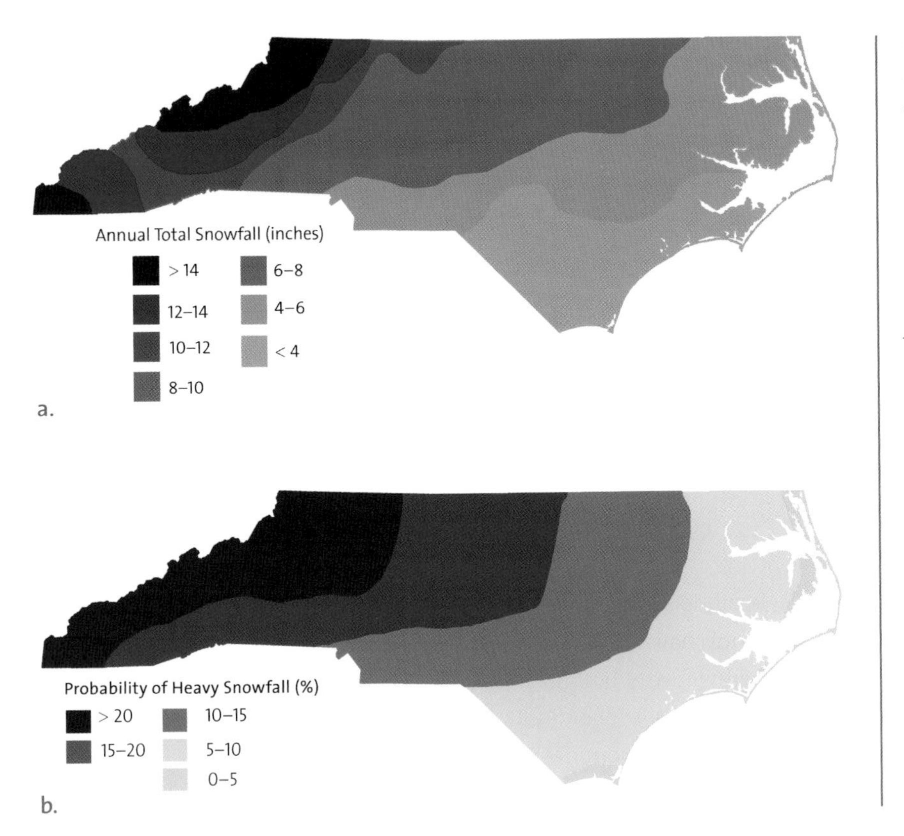

FIGURE 6.13.
(a) Average annual snowfall across North Carolina and (b) the probability that a storm will deposit 8" or more in any year. While there is a major decrease in both from west to east, the probabilities change much more slowly than the amounts, and there are significant differences within the Coastal Plain itself.

Spring

The Coastal Plain is the state's dominant agricultural region, and changes in the length of the growing season can have great economic significance. One measure of this length is the time between the last spring frost and the first fall freeze. There are great year-to-year variations in both of these dates, and in Chapter 2 we looked at the probabilities of frosts before or after specific dates. Also, in the mountain section of this chapter, we looked at the changes with altitude. But here we consider the possible long-term trends in the length of time between these dates, the growing season. It does appear that recently there has been a drift, more or less statewide, toward a longer growing season (see Fig. 6.14). For the Coastal Plain this involves earlier dates for the last spring frost and later dates for the first fall freeze. Year-to year variability and differences between nearby stations around the state, however, make specific statements and conclusions difficult.

Whether these changes are long-term effects associated with global warming

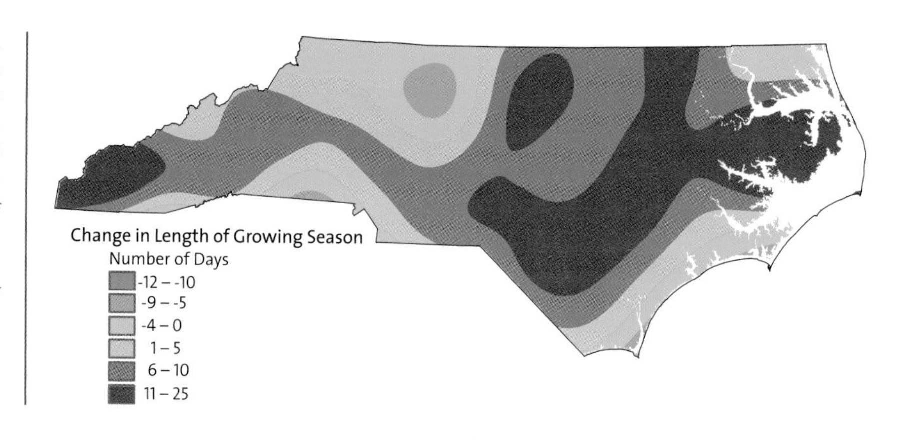

FIGURE 6.14. *Difference in the number of days between the long-term (1949-98) and short-term (1989-98) average length of the growing season across North Carolina. Much of the northwest seems to have had a shorter growing season in the most recent decade, as do several smaller areas. For much of the state the season has been getting longer, although there is no clear explanation for the pattern. (data courtesy of R. Boyles, North Carolina State Climate Office)*

or short-term fluctuations in our climate record is not clear. There is no ready explanation for the spatial pattern of the changes. Nor is it clear what meteorological or climatological processes are driving them. So, unfortunately, we cannot really use these results to predict the length of future growing seasons. All we can say is that for most of the state the growing season is now somewhat longer than it used to be.

Summer

Although Summer on the Coastal Plain is known for its heat, the beach has the most moderate conditions, and there is a rapid change as we leave the coast. We can show this using a set of stations in a line from the coast to the Sandhills near the Cape Fear River (see Fig. 6.15). For beach conditions, we have to use the Cape Hatteras observations, since this station is the only one truly close to the beach. Southport is somewhat sheltered from the full oceanic influence by Bald Head Island, while Wilmington Airport is about six miles inland, well away from the main oceanic effect. The other stations are successively farther inland, with Fayetteville and Pope Air Force Base being on the Sandhills rather than the Coastal Plain proper. There is a small increase in the monthly average temperatures for July as we move away from the beach to Wilmington airport, but there is little change after that. But the averages mask some major differences in day-to-day conditions. At the beach, very few July days have temperatures above 90°F, and equally few have nighttime temperatures below 60°F. The numbers for both hot days and cold nights increase inland and are particularly marked in the Sandhills. So the averages do not change very much, but the daily extremes in the summer are much greater on the inland portions of the Coastal Plain.

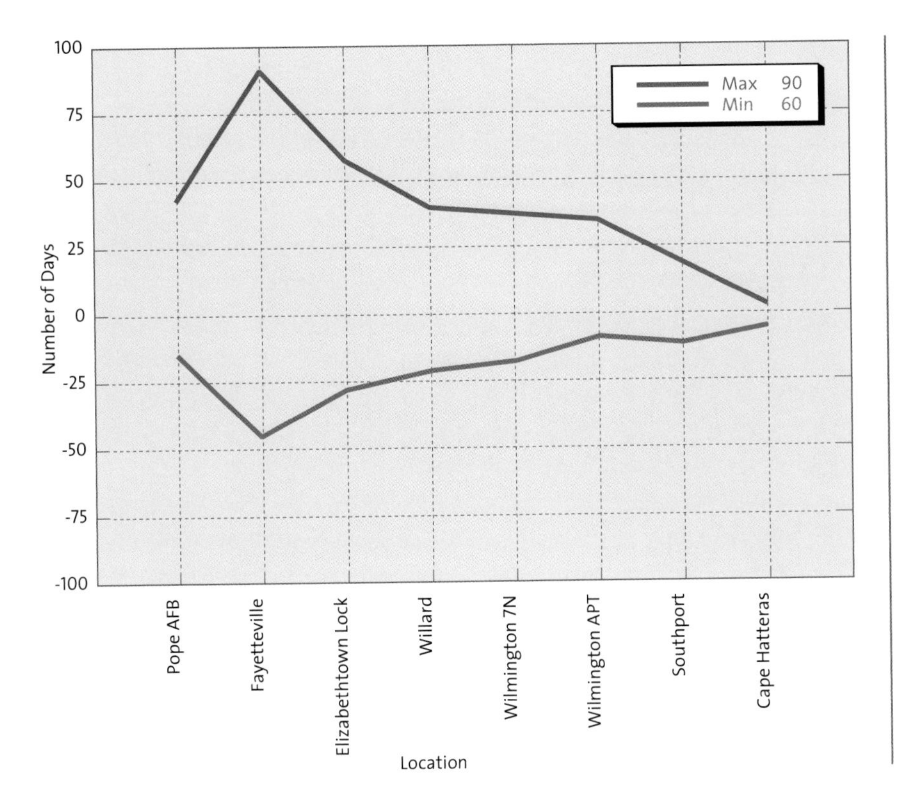

FIGURE 6.15. *Number of days each summer with maximum temperature above 90°F and minimum below 60°F for a group of stations roughly making a transect from the Sandhills to the coast along the line of the Cape Fear River. The number of days with minimum temperature below 60°F is given as a negative number to facilitate comparison.*

Fall

The foliage of the Coastal Plain changes color in the fall as surely as does that in the mountains, and for the same reasons (see Fig. 6.16). Indeed, the slow change of the seasons is the dominant weather feature for many weeks most years. But for most of us the most prominent atmospheric concern for a Coastal Plain fall is the threat of hurricanes. We normally think of these storms as violent and destructive forces. There is, however, another and more beneficial aspect for some storms. They can spread much needed rain over large areas of the Coastal Plain late in the growing season when water is needed but supplies are often low. After a hurricane crossed the Coastal Plain in late August 1924, the National Weather Service's *Monthly Weather Review* simply commented, "improved the crops." This beneficial aspect is particularly the case when the Bermuda high pressure region has dominated our weather for an extended period of time. Then rainfall from thunderstorms has been suppressed, and the rain-bearing wave cyclones have been pushed to our north. Drought is then an impending or actual concern. A hurricane passing through the state has the power to move the Bermuda high to

FIGURE 6.16. *Maritime tropical air and a few small cumulus clouds on a fall day in Camden County. The color of the foliage suggests that there has already been at least one instance when cold continental polar air has invaded.*

the east. So the hurricane not only brings rain directly but can also rearrange the atmospheric circulation so that wave cyclones and their rains also return.

The fall weather on North Carolina's Coastal Plain is, therefore, in one sense typical of our other regions and seasons. There are some spectacular, awesome, or fearful weather events embedded within an ever changing stream of weather. Sometimes the consequences are what we would expect; sometimes they are not so obvious. The components of the weather stream itself—the hurricanes and wave cyclones and fronts, along with the tornadoes and air masses—are well known, but their mix changes from season to season and year to year. Sometimes they are predictable a few days in advance; sometimes we are not sure what will happen even a few minutes ahead. But they all work together to produce the tremendous variety of our weather. They create our climate with four distinctly different seasons, and they help to produce the unique weather characteristics of each of the three major regions of our state.

chapter 7

Forecasting the Weather

In this chapter we consider primarily forecasts and forecasting, from the initial observations to how you can tailor a specific forecast to meet your own needs. We emphasize the daily weather forecast but also look at predictions for longer periods ahead, concluding with a section reviewing global climate change and the future of North Carolina's climate.

Observing Our Atmosphere

Before we can forecast the future of the atmosphere, we need to know its current state. We have mentioned some common instruments that we might find at a local observing station (see Appendix B) and satellite and radar observations. All of these observations have to be combined and compared, so here our concern is with the networks of stations working together to ensure that we know the current state of our weather.

WEATHER OBSERVING NETWORKS

A network is simply a group of stations taking the same kind of observations in the same way at the same time. Because of the uniformity, it is possible to make comparisons between observations. We can, for example, use the results of a particular network of thermometers to make a map showing the temperature variations across our state, confident that those variations are true weather differences. If we included observations from another network that used different instruments, we could never be sure that the map showed variations caused by the weather and not by the difference in instruments. For the same reasons, we maintain networks for

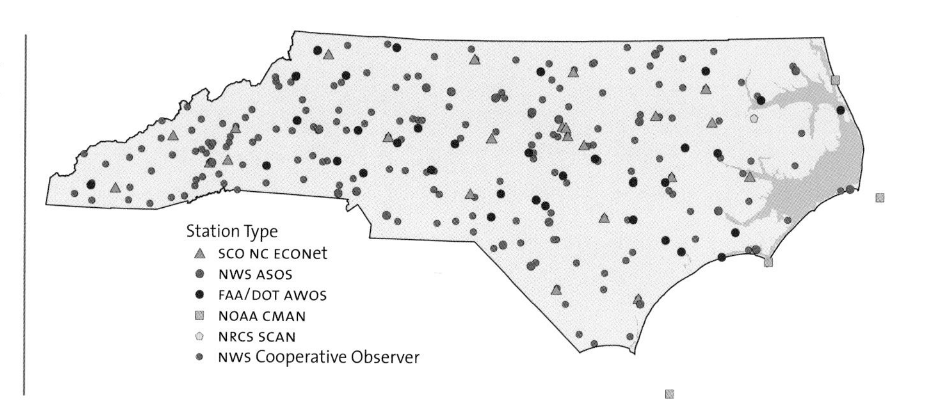

FIGURE 7.1. *Distribution of the stations of the major observing networks in North Carolina. The types of stations are indicated in Table 7.1.*

a long time, so that we can reliably determine whether the climate is changing or whether a particular event really was record breaking.

The first surface-based observing networks were established more than 100 years ago. These early stations measured temperature and precipitation, mainly to support agriculture. Since then the number of stations, the number of elements measured, and the use to which the information is put have increased considerably. Now there are a series of networks, some national and some local, observing our weather (see Fig. 7.1 and Table 7.1).

Stations in the federal Automated Surface Observing System (ASOS) network observe the wide range of elements needed for daily weather forecasting. As the network name implies, these stations operate continuously and automatically, reporting results once per hour directly to the National Weather Service (NWS). These results are also readily available to everyone on the Internet (see Appendix A). Most of the stations in this network are at airports, partly for historical reasons and partly because of the direct needs of the aviation community. Many of the ASOS stations in North Carolina were established during the 1990s. Prior to that, there were only six sites monitoring weather twenty-four hours per day, seven days a week: at Asheville, Charlotte, Greensboro–Winston Salem, Raleigh-Durham, Wilmington, and—representing the ocean environment—Cape Hatteras (see Fig. 7.2). As indicated in earlier chapters, at present we usually have to use these long-term stations for our climatology. We look forward to the time when the other stations have a long enough record for them to be used.

Two of our ASOS stations, Greensboro and Newport/Morehead City, also launch upper-air balloons called *radio-sondes*. Under the balloon is an instrument package that takes temperature, pressure, humidity, and wind observations as it ascends. The readings are radioed to the ground station. The ascending balloon

TABLE 7.1. Major Observing Networks Applicable to North Carolina

NWS ASOS and **AWOS**: These are the First Order stations making hourly observations, around the clock, of a wide range of elements. The long-term stations, recently automated, became the ASOS network. Newer stations, generally measuring fewer elements, are the AWOS stations.

NWS Cooperative Network: Daily total precipitation and usually daily maximum and minimum temperature observations are made by volunteers at a wide variety of locations around the state and the rest of the nation.

SCO NC ECONet: A statewide network of automatically reporting stations with a range of sensors, originally directed to agricultural use but now having a variety of uses, including emergency management.

Special Purpose: Many public utilities, government departments, and commercial firms make their own measurements for their own special needs. Parts of two national government networks, the CMAN buoys for marine interests and the SCAN station for agriculture are shown in Figure 7.1.

Local (educational/TV/governmental): Many media, local governments, or schools run their own networks, usually for a small area, for their immediate needs.

Note: Station locations are indicated in Figure 7.1.

expands and eventually, usually several miles above the surface and after a flight time of an hour or so, bursts. The package returns to earth on a parachute. Many unidentified flying objects have turned out to be radio-sonde balloons. The results of the ascent are used to give information about the vertical structure of the atmosphere, stability conditions, and the possibility of cloud formation. Our stations are part of an international network of sites, all of which launch their balloons at noon and midnight Greenwich Mean Time (7:00 A.M. and 7:00 P.M. Eastern Standard Time).

The other federal network is the Cooperative Network, which takes daily observations of maximum and minimum temperature and total precipitation. This is run by volunteers who cooperate with the NWS to ensure that the observations are taken in a standardized way at all stations. Currently there are about 150 active stations in North Carolina (see Fig. 7.1). Many of these have been in operation for decades—some for more than a century—and they are one of our best sources for information about climate and climate change. The climate maps in earlier chapters came from observations made by this network. Many local media also use it to report the local weather and compare it to the long-term normals.

The North Carolina ECONet (Environmental and Climatic Observation Net-

FIGURE 7.2. *National Weather Service ASOS station at Cape Hatteras. All instruments observe automatically and transmit the values to NWS headquarters. Originally located right at the cape, when it represented the closest we could get to a surface station over the ocean, it has now been relocated to Billy Mitchell Airfield, on the coast a couple of miles west of the cape.*

work), operated by the State Climate Office of North Carolina, combines a network of instruments taking measurements for agriculture, air pollution, and emergency management purposes with the ASOS network to create a single system to meet the special needs of the state. For example, all the observations are used in mesoscale models to provide detailed weather forecasts for hazardous weather.

In addition to these statewide surface networks, there are numerous special networks. Some are privately owned and make proprietary observations for specific needs. Most power companies, for example, monitor the location of lightning strikes. Many TV meteorologists have a network of observers scattered around the viewing area; sometimes they make routine measurements, and sometimes they act as "spotters" for various weather events.

The *Doppler radars* operated by the NWS also constitute a network providing complete coverage for the whole nation. This network ensures, for example, that the development and movement of a storm system can be tracked and forecasts can be issued even when the disturbance passes out of the range of one radar and into that of another. A radar is an active sensor that shoots pulses of electromagnetic energy, like a light beam but at a nonvisible pulse rate, from a rotating antenna

FIGURE 7.3. *Radar dome at Clayton, Johnson County, used by the National Weather Service station at Raleigh. The dome primarily protects the instrument and the electronics from the weather.*

(Fig. 7.3). When a pulse hits an obstacle—such as a cloud droplet or a raindrop—it is bounced back to a detector at the antenna. The time between emission and detection is measured and, because the energy travels at the speed of light, converted to a distance. As the antenna rotates, a map of reflections is created to give the familiar radar image (Fig. 7.4). The angle of the emitted beam can be varied so that the instrument can examine any level that seems especially active or interesting and provide a three-dimensional picture of the atmosphere. In the Doppler radar the energy is emitted at a wavelength specially chosen to be reflected by raindrops. However, as the pulse hits the moving water drop, its wavelength is changed slightly. If the drops are moving toward the instrument, the returning energy has a shorter wavelength than that emitted. The difference depends on the speed of the drop. Similarly, drops moving away return longer wavelengths. Detection of the differences allows us to determine the raindrop movement. This wavelength-changing effect of motion is known as the Doppler effect, named for Christian Doppler, the Austrian scientist who first identified it in the 1840s. Although the effect originally referred to light, the classic example has always used sound. In 1845 trained musicians recorded the changing pitch of a railroad engine's whistle

FIGURE 7.4. *Radar image from the Clayton radar, for March 20, 1998. Three major storm cells were present. The southeastern one spawned a tornado (identified by the "hook echo," the red hook seen in extreme south of the red area) near Garner. The bright white area in the western cell indicated hail, although this was high in the cloud, so it does not mean that hail reached the ground. (image courtesy of the National Weather Service, Raleigh)*

as it passed by; the high pitch (frequency) of the approaching train is replaced by a lower pitch as the engine disappears into the distance.

On a global scale, there is a network of *satellites*. The satellite that sends the pictures we see most often is in an orbit "parked" above the equator at 75°w. It is therefore south and slightly east of us and gives continuous coverage of our area. Other satellites at different locations above the equator cover other parts of the earth. There is also a network of satellites that orbit the earth so that they cross the poles twice a day. These are at a lower altitude, and twice a day they give a detailed look at our atmosphere. Satellites actually detect radiation upwelling from below. For cloud pictures, they sense the visible light reflected by the clouds or the ground. An infrared sensor, in contrast, detects the amount of energy emitted by the earth or the clouds below the satellite. This amount can be readily translated into temperature. The resultant image is usually color enhanced during processing so that the picture is more readily comprehended. Other sensors onboard a satellite allow us to determine the temperature structure of the atmosphere and help in forecasting cloud formation and precipitation. Yet other sensors detect the amount of water vapor in the atmosphere. Movement of water vapor can be used as a tracer for the upper level winds that ultimately steer surface weather systems.

THE PROBLEM OF MAINTAINING OBSERVING NETWORKS

So far we have simply said that a network is a series of observing stations all following the same procedures, allowing us to compare the weather from place to place and time to time. Maintaining a network, however, is often very difficult. Some problems are technical, such as keeping instruments and their shelters in good condition. One major problem for North Carolinians is maintaining a standard exposure at all stations. A station should be in the middle of a grassy area away from trees. This suggests a site in the middle of a cultivated field, but this makes access to the instrument inconvenient at the very least. A more accessible site near a residence usually has trees nearby, influencing the amount of sunshine onto an instrument shelter and the catch of a rain gauge. And trees grow continuously, increasing their impact with time. The value of placing observation stations at airports becomes obvious.

Even at airports, exposure often changes, sometimes in subtle ways. Since our state, and our nation as a whole, is becoming increasingly urbanized, our airports are constructing more buildings. Urban areas are warmer than nearby rural ones, so almost any station that has been observing for several decades is likely to have been influenced by this increasing urbanization and is becoming warmer. Sorting out this urban effect from true climate change is a challenge.

Once the observations are made, they must be recorded and archived for future use. This process may be as informal as writing down and tabulating your own readings or as complex as keeping the national climatic database. It is vital, however to maintain "quality control" of the data to ensure that they are a true record of the past weather. Instruments may malfunction, and observations may be written down incorrectly, lost in a power outage, or garbled in transmission. In general, errors seem to creep into the record almost like living organisms. Even small errors can lead to a false conclusion about a warming or cooling trend. Checks with earlier values and cross-checks with other stations nearby help to identify observations that might need correction.

Weather Forecasting in Midlatitudes

The most familiar type of weather forecast is the synoptic forecast seen every day on the TV weathercast. Here we look at how these forecasts are created and give some hints on how you can tailor the forecast for your own needs.

THE FORECAST GENERATION PROCESS

The production of any synoptic forecast involves six steps. The first two are primarily the responsibility of the NWS. Only a government agency has the resources needed to maintain a nationwide observing system, and only governments can make the international agreements needed to ensure that worldwide data are available. However, once the data are collected and made available, a variety of public and private concerns are involved in weather forecasting.

1. Observations of Current Conditions

Observations of the current situation are the starting point. We have considered these earlier, and here we can simply note that some network observations are used early in the process to get the national picture, and some are not required until very local conditions are of interest.

2. Transmission of Observations to Central Location for Analysis

Prior to the invention of the telegraph, weather forecasting as we know it was virtually impossible. Benjamin Franklin knew that the weather in Philadelphia was likely to affect New York a few hours later and Boston a few hours after that. But a rider on a horse could not "transmit" the information fast enough for the forecast to be of any use. Now, with modern electronic communications, almost instantaneous transmission is taken for granted. Within a few moments after synoptic observations are made, they have been received at the National Center for Environmental Prediction (in the Washington, D.C., suburbs).

3. Analysis of Current Conditions

In this step, the various observations arriving at the national center are used to create a series of surface and upper air maps. For most of us the surface map is probably the most useful (see Fig. 7.5). This preliminary analysis, in which fronts and isobars are added to the map, allows comparison of the current map with the preceding ones. This is vital, since continuity from one period to the next is a major characteristic of the atmosphere. Hurricanes do not suddenly appear, fully

FIGURE 7.5.
An analyzed map with the isobars plotted, fronts located, and centers of high and low pressure identified. The situation is twenty-four hours after that shown in Figure 4.2. The cold front is now well offshore, and all of North Carolina is dominated by cold northerly airstreams. The offshore cloud behind the front is probably the result of convection caused by the warm waters of the Gulf Stream creating instability in the cold air blowing off the land. (image courtesy of WRAL-TV)

formed, off the North Carolina coast. Indeed, a close inspection of the change in the positions of the fronts, isobars, and airflow patterns over the preceding few hours often suggests the conditions for the hours to come.

More detailed analyses of the current situation are performed automatically, using techniques based on mathematical models. These analyses emphasize the information needed for forecasting, such as changes in pressure, temperature, and humidity at individual locations on a regular grid on the surface of the earth and at various levels above it. In order to do the calculations, the models must make some simplifications and assumptions about the processes going on in the atmosphere. Since this simplification process introduces some uncertainty into the results, the NWS uses several different models, each of which makes different assumptions and produces slightly different results.

4. The National Prognosis

The national prognosis follows more or less seamlessly from the analysis; the same models are used for both. We use the term "prognosis" rather than "forecast" to indicate that it is a nationwide view of the future based primarily on model projections. There will be several different projections of the future, depending on the various models. The local forecaster must decide which model to use.

5. The Local Forecast

In this step the local forecaster evaluates and uses the national prognosis, adds local information, and creates our local weather forecast. Local experience and expertise is invaluable here. One model may seem to work best when a front is coming in from the west, while another may seem to catch the right mood when we have continental polar air dominating—and they need not be the same models for both Wilmington and Charlotte. Additionally, guidance is often provided by local models that are "nested" in the national models (see Fig. 7.6). These, usually called mesoscale models, incorporate local observations, take into account local surface features, and calculate local processes.

Everywhere in the nation seems to have several special situations that present a challenge to the forecaster. North Carolina is no exception, and we have already hinted at several anomalies. Perhaps foremost is the forecasting of the inland movement of hurricanes. The increasing accuracy of our forecasts of landfall times and locations has considerably reduced the loss of life along our coast. Now hurricane-related deaths occur mainly when individuals drown in flooded inland rivers, and concern has shifted to forecasting the inland movement of the storm and its rainfall amount. Perhaps equally important, and no less challenging, is determining the type of precipitation coming from a winter storm. Various relationships between air temperature and the form of precipitation have already been developed. But they turn out to be slightly different for Metrolina, the Triad, and the Triangle, and better understanding is needed to provide more reliable forecasts. Finally, less spectacular but often important is forecasting whether a sea breeze will form and, if it does, where and how vigorous it will be. As with the two other examples, national prognoses and mesoscale models can combine to increase the reliability of a local forecast.

There is a wide range of users of weather forecasts, from individuals deciding whether to carry an umbrella to farmers concerned with crop irrigation to utility companies projecting power needs. No single forecast can meet all the needs of all the users. The NWS, as the official federal forecast agency, has the very broad mission of providing forecasts for the preservation of lives and property. That, in fact, requires forecasting in all weather, and the forecast offices serving North Carolina provide the general "public" forecasts that are available every day. Threatening weather, of course, is a major concern, and these offices are responsible for issuing severe weather watches and warnings and for forecasting conditions that might lead to forest fires or river floods.

Many industries, such as commercial airlines, power companies, or agribusinesses, have very specific weather forecast needs. They, or private meteorologists

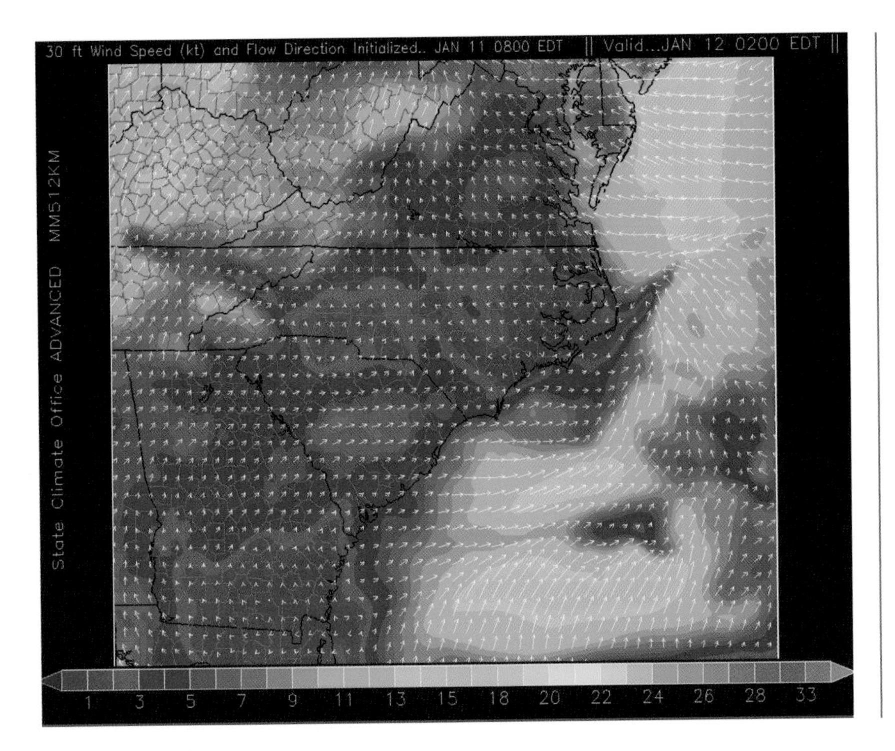

FIGURE 7.6. *Results of a mesoscale model simulation of the wind direction (indicated by the lines and arrows) and speed (color) showing strong offshore winds and light winds inland*

working with them, are likely to take NWS products, incorporate any observations of their own, and make the forecast. We rarely hear about this activity, but safe airline flights, efficient utility operations, or sufficient water in a reservoir depend on these forecasts. Similarly, many government agencies need specialized forecasts. For our state the ECONet observational network helps produce the forecasts required by agricultural, air quality, and emergency management agencies. Finally, for most of us the media are the prime source of weather information. Each media outlet follows a pattern similar to that above, although a particular outlet may relay the NWS forecast or subscribe to a specific forecasting service or make its own forecasts. Even if they use NWS prognoses, the media are normally expected to give the forecast the twist that makes it directly applicable to their viewing, listening, or reading audience and to use a format that renders the forecast as understandable and useful as possible (see Fig. 7.7).

6. Forecast Dissemination

The dissemination step has benefited tremendously from modern communication technology. With weather information available twenty-four hours a day on radio, TV, and the Internet, we can get the forecasts almost as soon as they are issued.

FIGURE 7.7. *Synoptic forecast issued by the media. This emphasizes the conditions of interest to most viewers in a form that can be quickly scanned and understood. It can also be updated easily and rapidly. (image courtesy of WRAL-TV)*

Several different but sometimes conflicting forecasts are usually available. Most of the time it is interesting to compare the local and national TV channels with the NWS forecast. Rarely is there much difference; but meteorology is not an exact science, and there is plenty of room for professional disagreement. Differences become significant only when life-threatening events loom. Hurricanes are the major concern (see Fig. 7.8). Their relatively slow evolution, the uncertainty of their tracks and intensity, and their potentially devastating effects mean that several equally plausible forecasts can be issued, especially several days in advance of a landfall. While discrepancies in forecasts provide a clear and realistic signal about our uncertainty, they can also cause problems for emergency management personnel or for anyone deciding whether to abandon a beach vacation.

REFINING FORECASTS FOR YOUR OWN NEEDS

These days there is a tremendous amount of weather information available (see Appendix A). For most of us most of the time, the daily newspaper and the radio or TV weather broadcast are all that we need. However, there are times when it is interesting, informative, or just plain fun to dig deeper and make your own forecast. On the Internet, virtually all of the information available to the professional

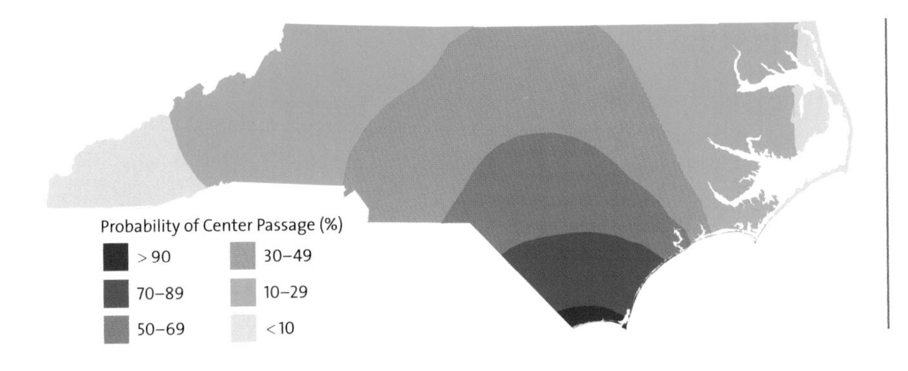

FIGURE 7.8. *A forecast showing the probability that the eye of a hurricane, in this case hurricane Fran in 1996, will pass overhead within the next forty-eight hours. All forecasts contain uncertainty, which increases the further we look into the future and decreases the more we understand the atmosphere. While our hurricane forecasts are getting more accurate, increased settlement on the coast means that evacuation takes longer, so we need forecasts farther in advance. The two effects are perhaps canceling each other out.*

weather forecaster is available for free. And their forecasts are also there, along with a discussion of why they arrived at the forecast.

Probably the first, and certainly the most obvious, thing to do when refining a forecast for your own needs is to look outside. The visual information often tells you immediately whether the forecast needs modification. For example, when a well-developed wave cyclone is passing through, your knowledge of the expected cloud and weather sequence tells you whether the system is moving as anticipated. Unfortunately, the times when cyclones look just like the ones shown in this book—or in any meteorology textbook—are rather rare. But it is immensely satisfying when they do occur in more or less recognizable form and you get the forecast right. In the same vein, if the forecast calls for a cloudless night and you see clouds building in the evening, you can use your knowledge of energy exchanges to suggest that overnight lows are likely to be higher than those forecast.

Knowledge of your local area, whether you are located in a frost hollow, a rain shadow, or a sandy area; sheltered by trees; or influenced by valley winds or sea breezes, should enable you to make modifications to the temperature forecast or, in some cases, the precipitation probabilities. This can certainly be done in an informal way, but an even more detailed refinement is possible if you make your own observations (see Appendix B) and compare them to those of the local NWS station over an extended period of time.

The most advanced method of refining the forecast is to use all the information that the NWS provides on its website to follow the sequence of steps used in the forecast development outlined above. However, this requires a great deal of training and much skill. I frequently look at the information to see how the weather is changing and how it is predicted to evolve. But this is much more out of interest in the weather than with any hope or desire to produce a more accurate forecast than the full-time professionals.

EQUATIONS, ANALOGUES, AND FOLK WISDOM

The preceding sections dealt with the modern methods of forecasting, using the current observations as the starting point for mathematically based estimates of how the laws of the atmosphere will operate in the near future. But other approaches are possible.

First, there is the "no skill" approach, which needs no meteorological knowledge whatsoever. Beyond a straightforward guess, the best of these is the "persistence" forecast: "Tomorrow will be the same as today." Anyone can give this. Slightly more sophisticated is the "persistence plus return to normal" forecast: "Tomorrow will be the same as today except that since today was warmer than normal, tomorrow will be a degree cooler than today." This requires only a knowledge of climatic normals. You can make this forecast yourself (see Appendix C for some normals). For summer temperatures in North Carolina, this approach works well. Our summers are dominated by hot, humid maritime tropical air day in and day out. An afternoon thunderstorm may make your maximum forecast a bit off, while the occasional wave cyclone passage may alter the picture completely. That is when the NWS comes into its own. But in winter, with its mix of maritime tropical and continental polar air, a persistence forecast can easily be off by 20° or even 40° if there is an overnight change in air mass.

We can use a similar "no skill but a bit of knowledge" approach for precipitation. The observational record indicates that for much of the state we have about 110 days with rain annually. Will it rain tomorrow? The record says there is a 110/365 chance—or that it rains about one day in three. So the odds are that it will not rain tomorrow. That is the no-skill forecast, and if we say it every day for a year, we should be right about 67 percent of the time. The NWS had better be (and certainly is) right more often.

Over the ages, people working outdoors, notably mariners and farmers, have developed methods of weather forecasting that have become part of our folk wisdom. In North Carolina we have imported some axioms from Europe and the northeastern United States and have also generated our own sayings. The familiar "red sky in the morning, sailor's (or shepherd's) warning," for example, refers to the passage of a wave cyclone. It was developed in Europe and is equally applicable to us. In theory, the morning sun low in the east shines across a nearly cloudless sky onto a cloud bank to the west of us. These western clouds, heading our way as part of a frontal system, reflect the sunlight, which is red because most of the blue light has already been scattered from the solar beam, to our eyes. At sunset the sun in the west shines on clouds to the east, and fair conditions are likely, since the front has already passed us. Unfortunately these folk sayings, even if they have some physical basis, do not prove to be a reliable guide to future weather. Our animal forecasters are equally unreliable. If the groundhog sees his shadow at noon on February 2, we are supposed to have a continuation of winter. However, there is no way to determine whether the groundhog is a good forecaster or not because no one can agree what it means to say that winter has been extended. Does it mean more snow lying longer, less sunshine, or low temperatures and, if so, how low and for how long? Our own woolly worms share the same problems, but they seem to have evaluated the climate during the preceding several weeks more carefully as they produced their various colors. But it has never been shown that their forecasts are any more accurate than those of the NWS.

More general weather statements, such as the prediction of an Indian summer or a January thaw, have a stronger basis when examined in light of our state's climatic record and our understanding of the response of the whole planetary atmosphere to the change in seasons. We do, for example, commonly have a warm period late in October before winter really sets in. But that period seems more marked in the Midwest than in North Carolina. Nevertheless, we can use the climate record to investigate a variety of potentially regular events and incorporate them into a forecast. In fact, the next section does just that. Along the same lines but with a broader scope, the forecasts given by the various almanacs for farmers use the climatic record as a guide. They may have their own special data and methods, and they certainly make for more interesting reading than most tables of climatic numbers.

Making Forecasts Using Past Observations

How early in the year can I plant my garden in Charlotte without risk of frost? We're having an outdoor wedding in June in Greensboro—is the weather better in the morning or in the afternoon? I'm planning a vacation in Asheville in September—what will the weather be like? If you are asking these questions a few months in advance of the event, the daily weather forecasts do not help much. Nor do the long-range forecasts, which tell about average conditions in the future, not the weather on a specific day. However, forecasts based on past observations can help answer these types of questions.

These forecasts are obtained by analyzing the past observations in such a way that it is possible to estimate the probability—or chance—of something occurring. So they are "probability forecasts." We mentioned them earlier as a form of no-skill forecast against which to compare the deterministic daily weather forecast. For our own individual long-range planning purposes, they offer a very flexible and useful method of generating information. We will look at some examples and then suggest ways in which we can create forecasts to meet our own particular needs.

EXAMPLES OF PROBABILITY FORECASTS

Scattered throughout earlier chapters there have been various probability forecasts. Here we can use precipitation data to give a set of examples progressing from simple to rather complex. We can start with the "chance of rain," which we calculated earlier as a no-skill forecast. We found that on a statewide annual basis, the chance of rain on any day was about one in three. We could easily take the observational record for a particular place and do the calculation specifically for that place. We could also divide the record into months and see whether the probability changed from month to month (using, for example, the data of Fig. 3.12). We could even do it for days of the week. As a final refinement, we might be able to look at the difference between morning and afternoon probabilities. By then we will have the information regarding the chance of rain for a June wedding, one of our original ideas.

We could continue to refine and expand our rainfall analyses in various ways. We have said nothing about variations from year to year or rainfall amounts or the chances of a sequence of dry days. The possibilities seem endless. The examples given, however, suggest some of the ways you can create your own probability forecasts.

MAKING AND USING A PROBABILITY FORECAST

These days it is very easy to obtain climatological data (see Appendix A) and make your own probability forecasts. The understanding of atmospheric processes and the use of climate information indicated in this book can provide guidance for appropriate types of forecasts and suggestions as to how valid or accurate they might be. But the type and detail of the forecasts you can make depend on your own interests, needs, and ingenuity. They also depend on your own statistical or computational skills and resources, which are well outside our concern here. Nevertheless, we can provide a few pointers to help you use the data wisely.

First, because it is easy to obtain data, it is easy to obtain lots and lots of data. Fifty years of daily data, easily downloaded, contains more than 18,000 observations. Making sense of that without using a computer—or even with a computer—

is difficult. So from the beginning, make sure you identify as closely as possible what you want to do.

Despite the seemingly boundless amount of information available, I sometimes think that it is a major law of probability forecasting that the right kind of data are never observed anywhere near any location where I want to make a forecast. So we usually have to make the forecast for one place and assume it applies at another or determine it for a series of places around our point of interest and interpolate the results. This is often easy for temperature and precipitation, where we have a dense station network. For other weather elements the networks are much less dense, and no observations are near, or even in areas climatically similar to, our forecast point. Then we must make major assumptions that may severely limit the accuracy and usefulness of a forecast. You will have noticed that we have avoided this problem as much as possible and have frequently used airport data without modification in this book.

Most of our examples have also assumed that there was no change in the climate over time. There is obviously day-to-day and year-to-year variability in the weather, but we have assumed that there has been no change in the long-term averages and thus no trend in the data. Usually there is a trend; but often it is small compared with the short-term variability, and we can safely ignore it. When it becomes important to consider trends, such as when we are looking at climate change, the determination of probabilities becomes more complicated, and more advanced statistics are needed to get reliable results. A final assumption we have made is that the data are complete and accurate. Most data from official sources should be reliable. However, most observational records have some missing data, and it is wise to check. The odd missing day scattered at random throughout the year is probably not a concern, but if July is missing every year, trying to calculate an annual mean might be a major problem.

All of these challenges, of course, can be overcome. Often this means using advanced statistics, detailed knowledge of atmospheric processes, and sometimes even mesoscale models. While I certainly do not want to discourage anyone from using the data to provide his or her own probability forecasts, I just want to emphasize that the forecasts must be developed with care. The methods suggested here are entirely suitable for general interest and guidance. But for detailed analysis or where a result is crucial for a particular application, much more sophisticated and refined methods are probably needed. Indeed, for most commercial applications I strongly recommend employing a professional meteorologist (see Appendix A).

Emerging Opportunities: Climate Forecasting

For many years the synoptic forecast and the probability forecast were the only means of looking into the future weather. Recently, however, we have begun to create temperature and precipitation forecasts for a season or more ahead and to predict the number of hurricanes likely to occur in the upcoming season. At present these are experimental extensions of ongoing research and do not apply specifically to North Carolina. However, as we gain experience and understanding, we are increasingly able to relate them directly to our state.

These advances are occurring because of our better understanding that processes in the ocean, in the atmosphere, and on the land are all closely linked. Some links can be explained by known physical processes. Others have been established by statistical means and, particularly those that are many miles apart, are known as *teleconnections*—remote relationships without clear physical bonds. In particular, the ocean tends to change much more slowly than the atmosphere, and by following the slow oceanic changes, we are able to forecast atmospheric changes several months into the future and often far from the ocean.

EL NIÑO AND THE SOUTHERN OSCILLATION

The area where the links between the ocean and the atmosphere, as well as the resulting teleconnections, are best known is in the tropical Pacific Ocean. The regular oceanic circulation there is sometimes disrupted by an event called El Niño, which occurs at the same time that the winds change in response to a pressure fluctuation known as the Southern Oscillation. In the most common situation the Pacific wind and ocean currents act together to produce a drift of surface water along the equator from near the west coast of South America to the east coast of Australia (see Fig. 7.9a). As the water pulls away from the South American coast, cold water rises to replace it, providing the nutrients for the schools of anchovies that feed there. At the same time, there is descending air and dry conditions. The westward-moving water is heated by the sun, and the warmest water is off Australia. This leads to an unstable atmosphere and clouds and rain. In addition, the constant flow of water leads to a higher sea level in the west than in the east.

From time to time, at an irregular interval that averages about seven years, this situation changes. The atmospheric circulation reverses, winds blow from Australia toward South America, and the warm water flows "down slope" eastward (see Fig. 7.9b). This eastward movement of warm water, seen by satellites, is the first sign that an El Niño event may be developing, although it can take months before the warm water reaches the Latin American coast. It may never reach there. When it does arrive, we have an El Niño event. The upwelling cold water is cut off, and the anchovies are deprived of nutrients. The descending air is replaced by ascending air, and dry conditions are replaced by wet ones—often exceptionally wet ones. Long ago this phenomenon provided needed rain for coastal agriculture, and the onset of El Niño was welcomed. Now it produces major flood problems as well as the collapse of the anchovy fisheries.

During an El Niño event the altered location of the warm water has a major impact on the distribution of the energy over the planet, influencing circulation patterns worldwide. Some clear teleconnections have been identified. We, and the southeastern United States as a whole, are likely to have a wetter winter than normal. Since we can track the Pacific water movement months before it arrives as El Niño, we can give a weather forecast months in advance.

SEASONAL CLIMATE FORECASTS

Teleconnections of the type associated with El Niño are being used to develop seasonal climate forecasts. Climate forecasts, unlike the familiar synoptic forecasts, deal with departures from normal, not with actual values, and they currently refer to conditions averaged over a three-month period. Hence they are seasonal forecasts. They are also probability forecasts and deal only with temperature and precipitation. The historical record is used to divide the three-month average into three ranges. The highest third of the values is labeled above normal; the middle third, near normal; and the lowest third, below normal. If there is no forecast information available, each category is equally likely; there is a 33.3 percent probability for each. When there is useful information, whether associated with El Niño, another teleconnection, or another slowly evolving feature, we can then tilt the probabilities in an appropriate direction (see Fig. 7.10).

These seasonal climate forecasts have a long *lead time*. The lead time is the difference between the date of issue of the forecast and the dates to which it applies. The farthest ahead a synoptic forecast predicts is about five days, so the maximum lead time is five days. The forecast in Figure 7.10 was issued in mid-January but does not apply until February, a lead time of two weeks. In early 2005 there were no strong teleconnection signals anywhere in the oceans and thus little guidance as to the conditions farther into the future. Even for the more immediate future shown in Figure 7.10, there is no indication for much of the nation what the conditions will be. For these regions the historical climatic information provides the best guide.

FIGURE 7.9. *Schematic diagram of the circulation in and above the tropical Pacific Ocean for (a) the most common situation (sometimes called La Niña) and (b) during an El Niño event*

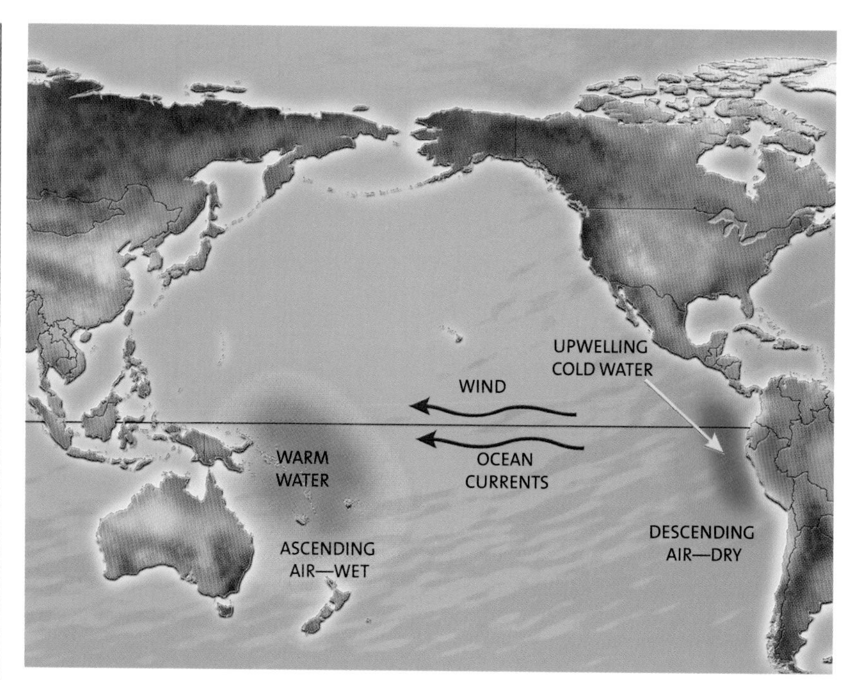

a.

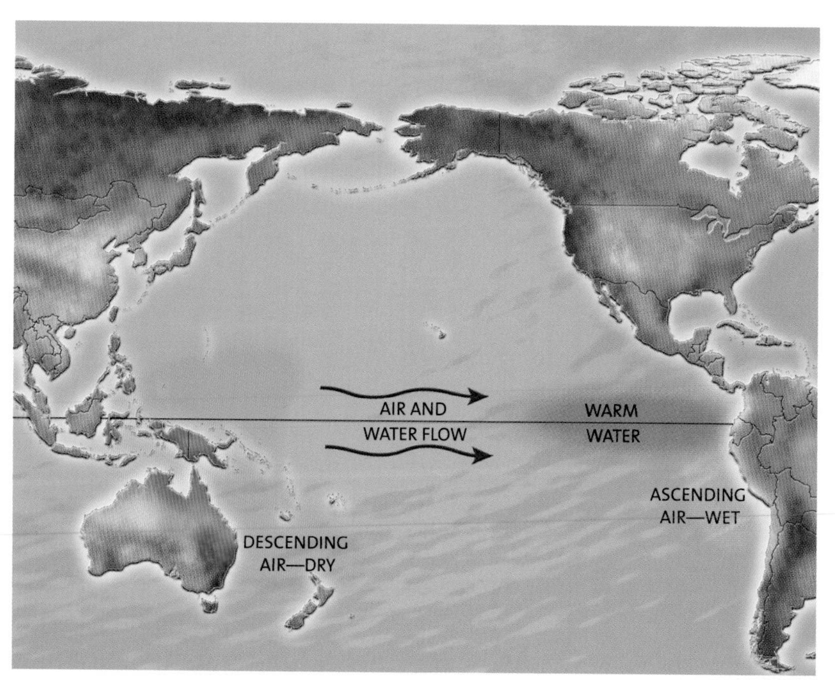

b.

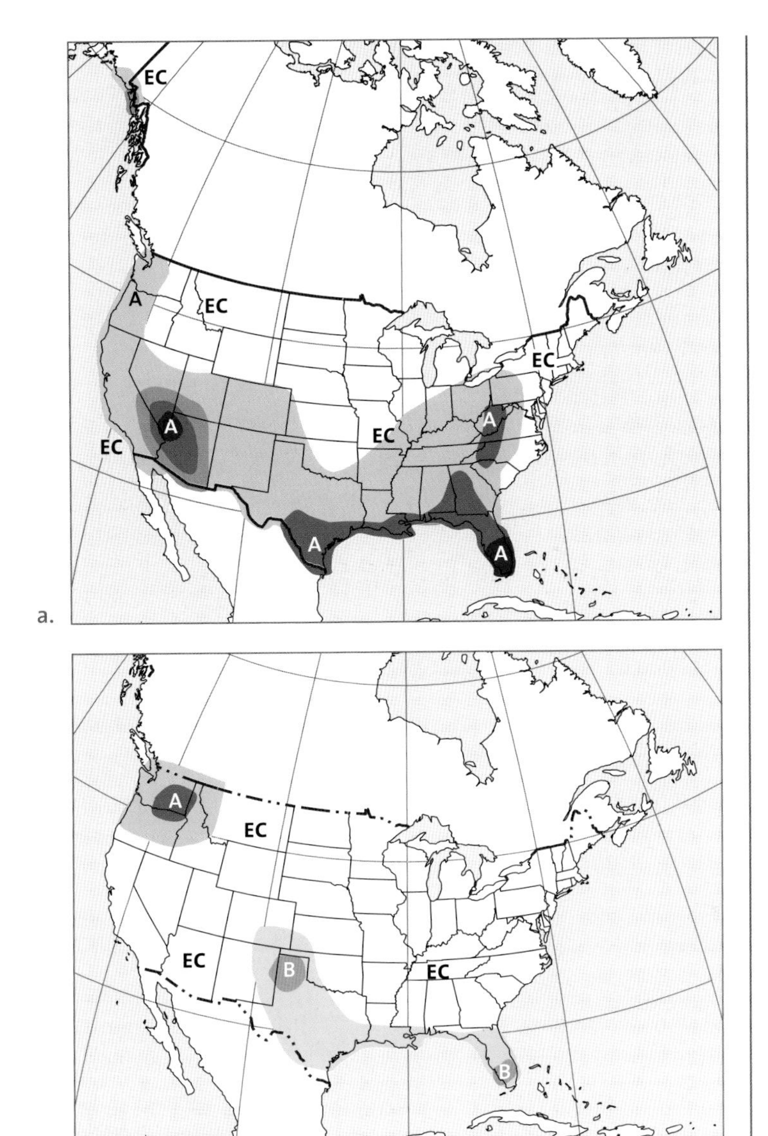

FIGURE 7.10. *Seasonal (a) temperature and (b) precipitation outlook for February-March-April 2005, issued January 20, 2005, by the NWS. Purple indicates where temperature or precipitation has a high probability of being above normal (A); blue, below normal (B). The deeper the color, the more confidence there is in the forecast. Much of the United States is forecast to have above-normal temperatures, with the Southwest, the gulf coast, southern Florida, and western North Carolina having a high probability of above-normal conditions. For precipitation in North Carolina there is not enough information to give a forecast other than having an equal chance (EC) of above-, near, or below-normal amounts. The gulf coast is forecast to be drier and the Pacific Northwest wetter than normal.*

ADVANCED HURRICANE FORECASTS

The availability of hurricane forecasts far in advance of the hurricane season has long been a dream of many North Carolina residents and a tantalizing objective for many weather forecasters. Increased investigations of teleconnections have led to progress toward that objective. Teleconnections between hurricane activity and a variety of early indicators have been established. Some, such as the sea surface temperatures in the Atlantic off the African coast, have strong links to processes known to affect hurricane formation. Other relations, such as the influence of the upper level conditions in the extreme south of the Indian Ocean, are almost purely statistical. Once a statistical link is found, however, it focuses the search for the physical connection, furthering our understanding and our forecasting ability.

These hurricane activity forecasts at present only suggest the likely number of hurricanes that will occur in the entire Atlantic Ocean area during the coming season. We are still far from predicting more than a few days in advance when or where individual hurricanes will make landfall. Indeed, all seasonal climate forecasts should be regarded as experimental. They do not have the broad, solid, scientific foundation that we expect from the familiar synoptic forecast. Nor is there long experience with making and evaluating them. Nevertheless, they have a strong scientific basis. Equally important, they can provide information that is better than a guess to many people and groups.

POSSIBLE USES OF CLIMATE FORECASTS

Because climate forecasts are still rather new and untested, the main users are industries long used to taking weather variability into account. Oil companies like to know as far in advance as possible whether the coming winter is going to be hotter or colder than normal. If we have a warm winter, we use less heating oil but travel more. So more crude oil is refined into gasoline rather than heating oil. Similarly, seed companies need to have information about the coming growing season so they can have adequate stocks of the appropriate seeds available for planting. Water companies with large reservoirs are beginning to use the forecasts to help control water supplies, minimizing the chance of a lake running low in a dry spell or a dam being overtopped in a wet period. The list of users and uses, however, is rapidly growing as experience and familiarity with the forecasts—and what they can and cannot do—increase. State agencies, local municipalities, and individual institutions are beginning to see opportunities to use forecasts to help with long-range planning.

Any forecast contains the chance that it will be wrong. Companies using seasonal forecasts commonly have large financial investments at stake. So with the new forecasts has come a new industry, that of *weather derivatives*, which uses the forecast information to hedge against risk on the commodity futures market. This industry is just beginning, but it appears to be rapidly changing and expanding the nature of atmospheric forecasting. Traditional concerns with saving life and property are evolving into situations in which multimillion-dollar decisions are involved.

The Challenge of Climate Change

The final section of this chapter on forecasting takes us to the very longest timescales and the issue of climate change. This is an area where the predictions cannot be produced in the same way as the daily weather forecasts. Instead, they are the end products of research investigations, and we must think of them as research results. As we gain better understanding and make better observations of the atmosphere, our research results should get better and more reliable. This means that they are also likely to change. The first modern estimates of climate change were made in the 1970s, when we were worried about potential global cooling. Now we anticipate global warming as a result of the enhanced greenhouse effect caused by the increased concentrations of carbon dioxide and similar gases in the atmosphere.

Most of our concern, and most research, has concentrated on the decadal-scale changes likely in approximately the next 100 years. This is the time span in which the consequences of the enhanced greenhouse effect are likely to be significant. It is also the timescale often involved when large, expensive, weather-sensitive structures ranging from power plants to dams are designed and constructed. A major problem with using these research results as forecasts is that we cannot simply sit around for a decade waiting to find out if we were right. At least that will not help us to improve our forecasting or understanding. One thing we can do, however, is to examine climates of the past and see whether we can "hindcast" them.

RECONSTRUCTING PAST CLIMATES

Various methods are available for reconstructing past climates. The instrumental record, which gives direct observations of the various components of climate, goes back no more than 200 years. Most information comes from the last 50 or 60 years. These short records do not capture all the variability of our past climates and, hence, cannot be a reliable guide to the future. To extend our knowledge, we must use *proxy* observations. These are observations of features—such as the date of flowering of a cherry tree—that vary with the weather and for which we have a long period of record. We use the period of instrumental observations to establish the exact relationship between the proxy variable and the weather, and then we use that relationship to infer the weather for the earlier period. There is a range of proxy variables involving many disciplines and well outside the traditional scope of meteorology and climatology. We mention selected examples here.

The most direct proxy method uses early manuscript records. In medieval Europe many monasteries kept records of annual crop yields and dates of grape harvests that can be related to the overall summer weather. They also often noted individual weather events, such as hail, to explain poor yields. In Asia the date the cherry trees blossom provides similar information often stretching back for centuries.

Several proxy methods are based on features that recur annually. Probably the best known is tree ring analysis, or *dendrochronology*. Trees produce annual growth rings that can be examined, without destroying the tree, by coring from the bark to the center and extracting a small, tubelike sample from the tree. The growth rings are sensitive to the weather, and trees that grow close to the limits of their climatic range show the greatest growth variation from year to year. Much early dendrochronological work was performed in the U.S. desert southwest, where water is the major concern for most trees. Here a thick growth ring indicated a wet year. For North Carolina, water is rarely in short supply, but many of our tree species are sensitive to temperature. A cold year gives a thin growth ring (see Fig. 7.11). By combining both methods and coring many trees, we can develop a picture of the climate several centuries into the past.

Ice core analysis uses a similar technique but can take us farther into the past. Layers in ice develop because of differences in snow accumulation rate with the seasons. They help us describe the climate because the chemical composition of the air trapped by the snow as it lands depends on the temperature. Thus we can obtain a temperature record. But these records are obviously restricted to cold latitudes. More widely available are records from the sediments that collect in an annual cycle on the beds of lakes and oceans or in sheltered places, especially caves, on the land. Botanical, archaeological, geological, and oceanographic techniques can be used to extract the climate information.

TRENDS IN CLIMATE FOR NORTH CAROLINA AND THE WORLD

Using these proxy techniques, we are getting a clearer and clearer picture of the past climates of our planet. The farther back in time we go, the fewer the obser-

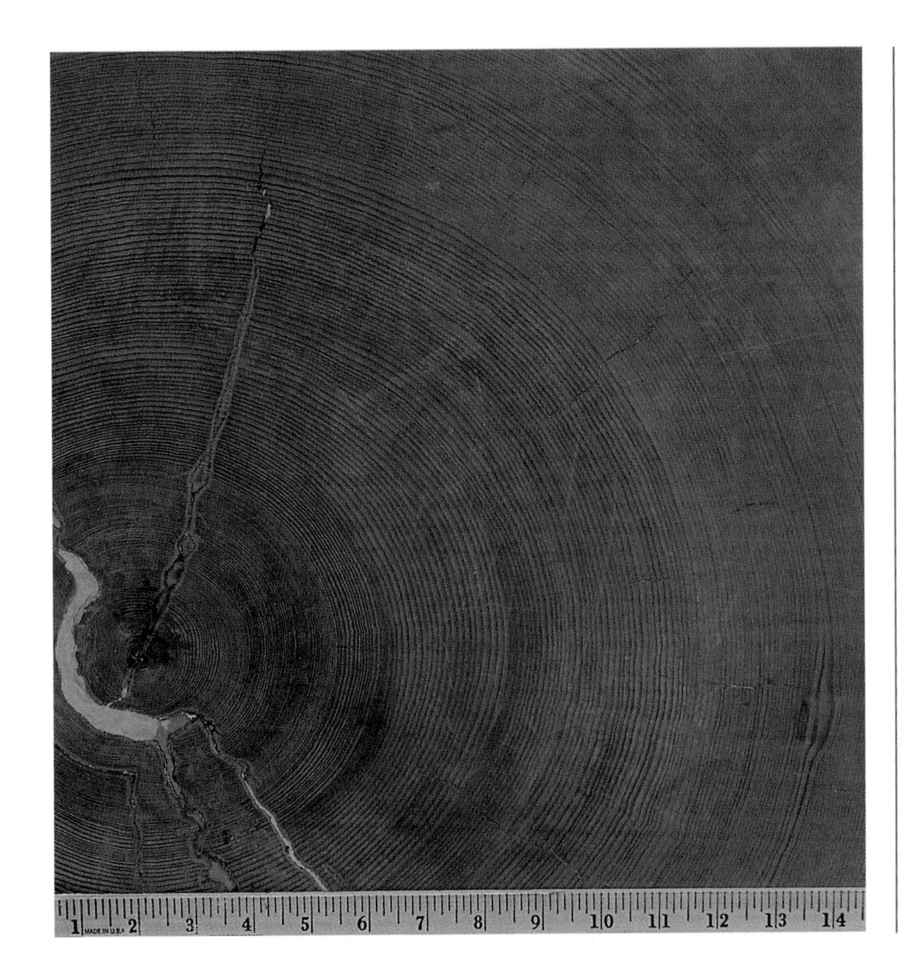

FIGURE 7.11. *A portion of the growth rings of a North Carolina tree more than 800 years old. At the center are a series of thin rings made by the young tree. Thereafter, the higher the temperature, the wider the ring. Until about 6" on the ruler there had been a warm period. Then came a band of thin rings, lasting to about 8" and indicating a cold period. This was followed by a warming; then around 11" another cold period set in, lasting for most of the rest of the life of the tree. These narrow rings correspond to the Little Ice Age in North Carolina that started not long before Columbus discovered America.*

vations available and the more general the results. For most of the history of the earth, we cannot separate North Carolina's climate from that of the rest of the planet. Globally there have been warm and cold periods lasting millions of years; there were times when the planet was virtually ice free and times when it was almost completely glaciated. During the last 1 million years or so the world has been in a cold phase, with several periods of glacial advance and retreat. Currently the ice sheets are naturally in a period of retreat, but we can expect them to advance again in the next few thousand years. Indeed, the climate is constantly changing, from the familiar daily cycle superimposed on annual events to cycles that can last for millions of years, and which we do not understand very well.

For our state, probably the earliest trend with any direct impact on us now is the warming that started about 15,000 years ago. This was the end of the latest Ice

Age. We were considerably colder than we are at present, although not sufficiently cold to be permanently covered in snow and ice. That condition was not present much closer than Pennsylvania, except possibly on some mountain peaks. But we did have vegetation that adapted to this much colder climate. As the whole planet warmed, the glaciers and polar ice caps started to melt, and the level of the sea rose. Parts of the Coastal Plain were flooded. The warming was not continuous or uniform. There were times when cooling, lasting perhaps 2,000 years, dominated. As the climate fluctuated, so did the sea level. Geologists have identified several old shorelines on the Coastal Plain that even now have an effect on the drainage patterns of the area. The Little Ice Age between about 1400 and 1850 A.D. was a cold period with low sea levels, and since then there has been a slow, natural rise in the level of the sea. This is clearly associated with the general worldwide warming over the last century or so indicated by our observational information (see Fig. 7.12a).

The southeastern United States, however, is one of the few areas on earth where there has been no major warming over the twentieth century (southeast China is the other large region). We in North Carolina have had fluctuations throughout that period (see Fig. 7.12a). There was a general warming from the beginning of the century until about 1940. This was followed by a fairly rapid cooling until the end of the 1960s, with a warming trend thereafter. Overall there has been little change, with the year-to-year fluctuations overshadowing these slight trends. The situation for precipitation is similar but even more marked (see Fig. 7.12b). There has been no detectable trend, but there were great year-to-year fluctuations.

PREDICTIONS OF FUTURE CLIMATES

Predictions of future climate are made in a way similar to that of predicting tomorrow's weather. Climate models are, in essence, the same as the models used for weather forecasting. The climate models, however, must take into account some things that weather models can ignore. Seasonal changes in the amount of solar radiation received, for example, can be ignored when we are forecasting for a few days ahead; but cumulatively they have an effect that has an impact from one season to the next, and eventually they influence the whole climate system. Indeed, virtually all of the various processes indicated in this book as driving the climate must be understood and used when we try to determine the nature and causes of climate change. So numerous climate models have been developed, all somewhat different in their assumptions and detail. Most models these days can reproduce the major features of the past climate, giving us some confidence in

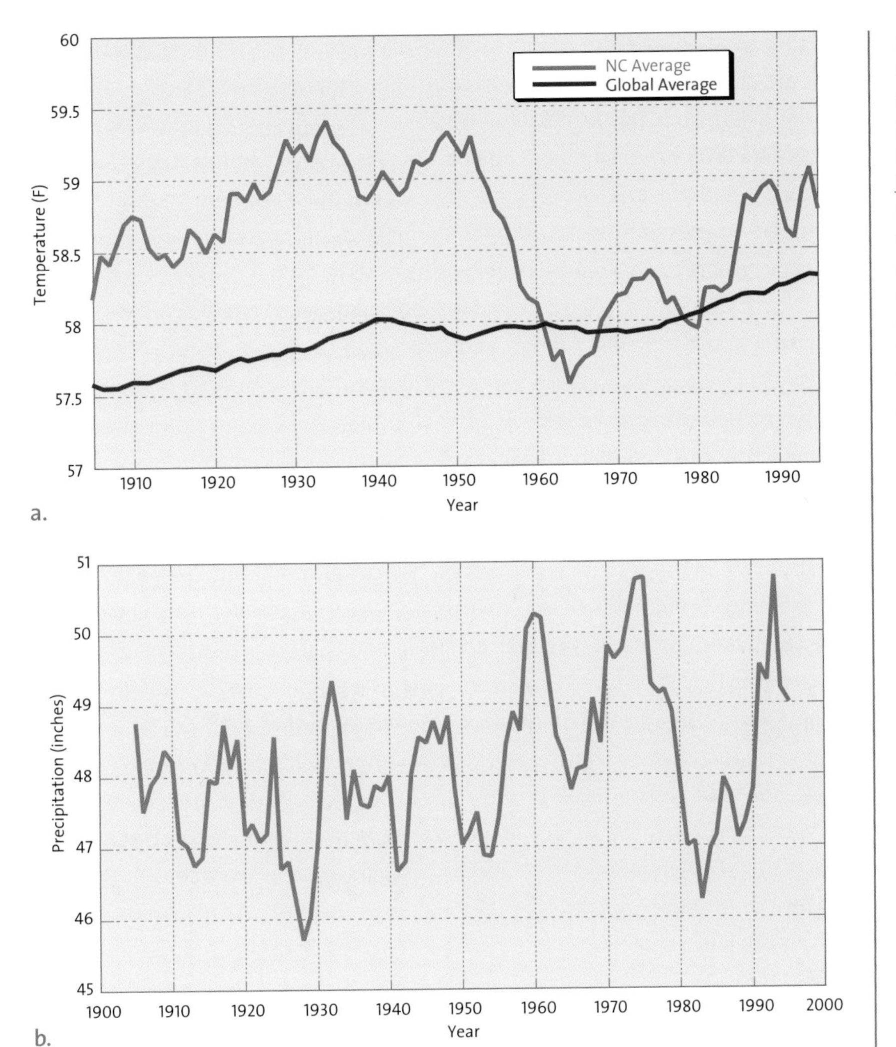

FIGURE 7.12. *Statewide average annual (a) temperature and (b) precipitation for North Carolina in the twentieth century. Consistent long-term changes are not apparent in either. The clear trend in global temperature is shown in part a.*

their ability to predict the future. But one major problem with hindcasting the recent period, where we are most confident about the climate data, is that the data already incorporate the role of human actions increasing the concentration of the greenhouse gases and altering the nature of the surface. Sorting out natural and human changes becomes very complicated. When we try to make predictions, one of the major causes of uncertainty is the amount that human activity will influence the surface and the atmosphere. Thus we must consider predictions in many fields, such as economics, politics, and law, that are far removed from climate.

Even when we make assumptions about this level of human activity, it is unlikely that we shall have the expertise necessary to provide reliable suggestions for the future climate specifically for North Carolina anytime soon. There continue to be major efforts to meet the challenges and uncertainties associated with climate change. The Intergovernmental Panel on Climate Change, an international group of scientists and government officials, issues periodic reports on the state of our knowledge. At present it is unwilling to do more than suggest possible future climates on a global scale, and the panel warns that the global climate is warming.

For North Carolina, it is certain that we shall continue to have variable weather. Droughts will occur at some times, and floods will happen at others. Some years will have hurricanes, and others will not. High and low temperature records will continue to be broken from time to time, and, occasionally, weather that seems to come close to the climatic normal will occur. Almost certainly the statewide averages of temperature or precipitation will change over the next century, but the amount is not certain. The sea level will probably continue to rise as the planet warms, but again the amount is not certain. So specific preparation for a particular climate change is not possible, since we do not know what that future climate will be. Nevertheless, preparations that acknowledge that variability will continue, that extremes may get a bit more extreme, that the frequency of hazardous weather events will differ in this century from the last, and that averages will change, is a prudent response to the challenge of weather, climate, and climate change. Overall, however, North Carolina will continue to have weather that is wonderful in its variety, whether the silence of a cloudless mountain morning after a winter snowstorm, the splendor of dogwoods blooming on a bright spring day in the Piedmont, or the drama of a hurricane approaching the coast on an autumn night. I hope this book has helped you to appreciate that weather more fully and live with it more wisely.

appendix A

Getting Information

There is a wealth of sources for data and information relating to the atmosphere. Some have been around for a long time, and some are still developing.

For weather forecasts, many sources are well established and familiar: the local TV and radio media, the Weather Channel and similar cable TV sources, newspapers, and of course, the National Weather Service. Many of these sources also give some climate information—long-term normals or extremes for a given date, for example—mainly to put the current weather into the context of what is usual or familiar for a particular place and time of year.

For climate information and for information about past weather events, there are numerous places to start, but there are three major public sources. For information about places in North Carolina and about the state as a whole (and probably the best starting place for any initial inquiries), see

State Climate Office of North Carolina
Website: <http://www.nc-climate.ncsu.edu>
Suite 240, Research III Building
1005 Capability Drive
Centennial Campus, Box 7236
North Carolina State University
Raleigh, NC 27695-7236
Phone: 919-515-3056, Fax: 919-515-1441
e-mail: sco@climate.ncsu.edu

For information about other southeastern states, see

The Southeast Regional Climate Center
Website: <http://water.dnr.state.sc.us/climate/sercc>
2221 Devine St., Suite 222
Columbia, SC 29205
Toll free: 1-866-845-1553
Phone: 803-734-9560, Fax: 803-734-9573
e-mail: sercc@dnr.state.sc.us

For national and much international information, see

The National Climatic Data Center
Website: <http://www.ncdc.noaa.gov>
Federal Building
151 Patton Ave.
Asheville, NC 28801-5001
Phone: 828-271-4800
Fax: 828-271-4876
e-mail: ncdc.info@noaa.gov

For international climate information, the single best starting point is

The World Meteorological Organization
Website: <http://www.wmo.ch/web-en/member.html>

This provides links to the official Meteorological Services in more than 100 different countries. Most provide some kind of climatic information, although the amount of information available, both online and by mail, varies tremendously from country to country. Some places charge for climate information. Not all sites are in English.

The range of Web-based resources is immense and is continually changing and expanding.

All three U.S. centers listed above have websites that contain a great deal of information on all aspects of climate and lots of data available for downloading. Much of it is free. They also have links to other weather-related sites.

For convenience I use the North Carolina State Climate Office website as my own Web starting page (I use this site partly out of loyalty to the office where I work part time, partly because I have some say in what is on the site, but primarily because it is exceptionally convenient). With only a few clicks I can readily get almost all the past and current North Carolina climate data I need in my professional work and all the current weather data and forecasts I like to have as someone simply interested in the weather. There is an option on the opening menu for "Related Resources" that gives links to other weather-related sites. We check and update this periodically and try to ensure the best access to sites on the Web that are not only the most useful but also are reliable, trustworthy, and accurate.

One set of links goes directly to the North Carolina offices of the National Weather Service. Each local office gives a range of local information but always includes current conditions, synoptic forecasts, and explanations of those forecasts. Links to the models used to make the forecasts are also provided. Often there are analyses of recent notable weather events on these sites.

This wealth of information allows anyone to make his or her own forecast or determine weather and climate trends. However, when information is needed for professional rather than hobby purposes, it is not necessarily desirable to do it yourself. There are numerous meteorological consultants available for this situation. The professional organization for meteorologists, the American Meteorological Society, has established the Certified Consulting Meteorologist program to identify individuals who have met established standards in the field. Consultants frequently advertise in the Bulletin of the American Meteorological Society, and a list of certified consultants is available from the society.

American Meteorological Society
Website: <http://www.ametsoc.org/memb/ccm/ccmhome.html>
45 Beacon St.
Boston, MA 02108-3693
Phone: 617-227-2425
Fax: 617-742-8718

appendix B

A Practical Guide to Observing the Weather

There is an old, familiar saying, "Everyone talks about the weather, but no one does anything about it." Maybe we cannot do much about it, but we can make observations and take measurements related to the weather where we live. Making your own observations requires an investment in time and money. So first we suggest ways to use observations, and then we review some of the options for choosing particular instruments.

Why Take Your Own Measurements?

What you do with your observations depends on your needs and interests. Most people like to keep track of their day-to-day observations and compare them to the local official National Weather Service (NWS) readings. Comparing our home-place with the local "official" station often gives us bragging rights about being colder, hotter, or wetter than the official station. These days comparisons with local or even remote stations are easy, since there is so much information available on the World Wide Web (see Appendix A). Even without that resource, the local newspaper usually records yesterday's weather at the local station, which allows you to compare results. Such comparisons are always interesting, but they can also be very instructive not only about weather in general but also about the weather and climate at your own observation site.

We can use temperature observations as an example of some simple comparisons and the questions such comparisons often raise. As the length of your record increases day by day, it is likely that some patterns will emerge. You may be consistently warmer than the NWS. (For many years most of my meteorological friends taking observations in the Chapel Hill area knew that Raleigh-Durham airport was usually colder than Chapel Hill.) But the amount of the difference will vary from day to day, and some days you may be warmer. Does the difference vary with season or with wind direction? Is the difference more marked when it is cloudy or when it is sunny? Does it reflect daytime high temperatures or nighttime lows or both? Why is there a difference? Are you in a windy situation or a frost hollow? The questions soon seem endless.

On a more practical note, you could track your electrical energy usage against the temperature at your home to see how efficient your house is. Many of us think about this on a monthly basis when the bill comes. We can then compare the monthly temperatures with the amount of energy we have used. But it is also possible to read the meter ourselves on a daily basis and track usage precisely. You can also track the length of the growing season or the date of first or last frost. Or you can translate the general forecast issued by the NWS or your local meteorologist for your own specific situation.

Comparisons involving temperature or humidity are usually fairly easy to do, and often that is the case for wind speed or direction. But rainfall comparisons are a very different matter. You can easily get an inch of rain while the NWS a few miles away gets none. Is your gauge correct? It probably is, since a summer thunderstorm is great for giving that kind of spatial variability. Not until you add up rainfall amounts over a month or a year do you begin to get useful comparisons. Annual totals are usually assumed to be less variable than daily amounts. So after a year, your values should not be too different from those of the NWS. Again, that may not be true—either because of problems with your location and instruments or because there has been a real difference in the weather. With rainfall it is often difficult to know why your values differ from the official readings, but instruments are rarely completely wrong. Making a comparison with several stations in your area should suggest the kind of variability you can expect from place to place and where you fit in that scheme.

Selecting and Siting Instruments

Remember that differences between your site and the local NWS station are going to arise from two effects. First, you expect the weather to be different in different places, which is why you take the observations anyway. But second, different instruments located close together, or similar instruments in slightly different locations, will give different readings. It is expensive, if not impossible, to duplicate the NWS instruments, and few of us have the center of a level airfield, or any kind of field, available for our instrument location.

Weather instruments commercially available range in type and price from the relatively simple home thermometer and barometer combination with direct-reading dials, often designed at least partly as a decorative home accent and costing a few dollars, to fully instrumented, professional quality weather stations costing tens of thousands of dollars. Few of us need, can really use, or can afford the top-end equipment. Usually, however, the more you pay, the more accurate, reliable, and rugged are the instruments. Costs increase because all instruments have to be calibrated—compared and modified to match instruments that are known to be accurate. Calibrating a large group of thermometers all together at only two different temperatures is obviously less costly and accurate than treating each thermometer individually and calibrating it at ten different temperatures. The first group will not be wrong; they will simply not be as accurate. For most of us, accuracy is probably less vital than cost. My only advice is to buy the best you can afford and concentrate on enjoying taking the measurements.

Recent advances in electronics mean that almost all instruments are continuously operational and can be read, and their readings recorded, indoors. Gone are the days when the dedicated observer had to go outside in a blinding snowstorm at the crack of dawn to peer at the etched marks on a mercury-in-glass thermometer and try to write down the observation. Now the instruments may be outside, but the dials that show the values or the computers that record them automatically are comfortably indoors.

The ideal site for all meteorological instruments is in the middle of a flat piece of land, probably several acres in extent, that has the same type of surface, preferably low-growing grass, throughout. There should be no obstacles to wind flow—such as trees or buildings—for a long distance, preferably several miles, in any direction. Even the NWS, using airports, has trouble meeting these requirements, and few of us can come close. This is especially true in much of North Carolina, with our abundant tree cover. So again, my advice is simply to do the best you can to get an open, flat, obstacle-free site. Bear in mind the character of your site. A rain gauge under a tree will give readings different from that at the local NWS station, so you can compare how much rain your tree catches. At the same time, your observations are probably closer to the amount that actually reaches your garden than are those of the NWS.

Each weather element, of course, has to be measured by a separate instrument, and we can comment on the specific instruments for the most commonly measured elements.

RAINFALL

In essence, rainfall is measured by collecting rain in a container and measuring the depth that has been collected in a known interval of time. Indeed, most of the cooperative weather stations still use this method, observing once a day and using a rain gauge meeting the NWS criteria for orifice size, height above ground, and distance from upstanding obstacles (see Fig. AB.1). More continuous records are obtained by using tipping bucket gauges. Here the falling water is caught in the gauge and funneled into a small bucket. Rain collects in this until a known amount, usually 0.01", has accumulated. The bucket then tips, emptying the water and triggering a recording device. The process then starts again. An alternative approach, using pressure transducers to detect the rain collected in the gauge, has been developed, but at present it is not sufficiently accurate and rugged to meet the requirements of the NWS.

Rain gauges for home use probably come in a greater variety than any other weather instrument. A plastic wedge gauge available for a few dollars from a hardware or garden store will give a reasonable idea of the amount that falls, provided you put it on a post some distance above the ground so that it does not catch water splashing in as well as falling in. Fancier, more expensive rain gauges can replicate the NWS designs and may allow you to compare your amounts with the local NWS station more directly. But for rain, open exposure is more important than the type of gauge. Get as far as possible from obstacles to wind flow and driving rain.

FIGURE AB.1.
The two National Weather Service rain gauges at Elizabethtown Lock, Bladen County. The catch in the smaller gauge is measured once a day. The larger one is a recording rain gauge, and the amounts are recorded once every fifteen minutes. In most parts of heavily wooded North Carolina it is difficult to get an ideal open site. The location here is somewhat sheltered by the line of trees in the background, although the land is open in all other directions.

Even more difficult is the measurement of snowfall. Most desirable is a measurement of the depth accumulated in the middle of an open field free of drifts. Even then, unless it is still snowing, there will be some evaporative loss. Walking out to the site disturbs the surrounding snow, so measurement of the next day's snow amount is likely to be less accurate. The common alternative is to let the snow accumulate into a rain gauge, melt it, and measure the amount melted. While this gives the water equivalent of the snow, which is often of great concern to people involved in water supply, the measure is not very accurate.

TEMPERATURE

The electrical resistance of some metals varies with their temperature. This effect is used by modern thermometers, since the sensor can be placed in a suitable spot outside and the signal conducted to a convenient indoor location for recording. Since we want to measure the temperature of the air, the location and exposure of the sensor is as important as the sensor itself. The thermometer should be placed in a shelter that prevents direct sunlight from reaching the instrument, minimizes the impact of energy rising from the underlying surface, and allows free air circulation around the instrument. Currently, sensors are commonly encased in a white-painted, louvered housing, which has replaced the white "beehive" housing that for many years was the sure sign of a weather station.

HUMIDITY

The NWS uses a cooled mirror approach to humidity measurement. Condensation occurs on the mirror when it is cooled below its dew point. This is detected through alteration in the amount of light reflected by the mirror. As soon as the dew point is reached, cooling is replaced by heating until the condensation evaporates. Then it is cooled again. This cycling continues until a stable condition is achieved. A temperature sensor embedded in the surface of the mirror indicates the dew point. The entire sensor is usually housed in the same shield as the temperature sensor, and for the same reasons.

WIND SPEED AND DIRECTION

In the most common form of anemometer, wind speed is measured by determining the rotation rate of a spindle. In some versions this rotation is a response to a wind-driven propeller; in others a series of small cups are pushed around by the wind. In either case, the number of rotations in a fixed time interval is used to determine the speed.

Wind direction is measured by a wind vane (sometimes called a weather vane) that points into the wind.

Probably the most difficult aspect of wind measurement is getting the correct exposure. Nearby obstacles alter the speed and direction of wind, while in the lowest layers of the atmosphere, wind increases with height above the ground. The World Meteorological Organization has established a standard whereby all measuring instruments should be on a mast 10 meters (about 30 feet) high and approximately 100 meters from the nearest upstanding obstacle, such as a tree. Even the NWS at airport sites sometimes has difficulty meeting this criterion. Most of us are forced to set the instrument as high and as far as possible from obstacles. This often simply means that we use a convenient rooftop location. But differences in exposure between stations do make comparisons very difficult.

AIR PRESSURE

Air pressure has traditionally been measured by the height of the liquid column in a mercury barometer. Frequently the fragile glass column has been protected by an intricately carved wooden case, and for many people such a barometer has become a piece of furniture as well as a scientific instrument. For more meteorological use, aneroid barometers are used. These detect the amount of flexing of a partially evacuated metal container as the air pressure changes. This method avoids any use of mercury, a highly toxic substance, while the flexing can be recorded automatically. The flexing can also be linked to a pressure indicator dial. This device has also commonly been used as a home instrument, with various forecasts linked to various pressure tendencies.

RADIATION

Radiation sensors commonly operate by letting the radiation fall on an absorbent surface and measuring the resultant temperature increase. The surface needs to be kept clean, so it is usually protected by a transparent dome. For solar radiation, glass is used, since it transmits all appropriate wavelengths. There are also instruments that detect ultraviolet radiation, and it is even possible to measure terrestrial radiation. However, accurate measurements of radiation are very difficult to make and require fairly expensive and complex instruments. Further, there are very few NWS stations with which to compare measurements.

OTHER ELEMENTS

Other instruments are available for the determination of specific aspects of the weather and climate. Most are used in specialized situations. Visibility sensors, for example, may be important in areas particularly susceptible to fog. In other places a simple "eyeball" observation suffices, with fog being recorded when the observer can no longer see an object that is known to be a quarter-kilometer away. Cloud amount can also be measured or estimated by eye.

WEATHER STATIONS

Often "weather stations" are sold that measure a variety of elements using sensors chosen to give comparable readings or types of output. Like the individual instruments, they range from home systems to stations for scientific research sites. Prices vary accordingly. However, many of them take advantage of the included electronics to store historical information or make calculations. Many, for example, store data long enough to provide information on the maximum and minimum temperatures during the preceding twenty-four hours. Some also combine the temperature and humidity readings to determine the heat index. These additional features may be very important in making a purchasing decision.

Getting Started

To get started, decide what you want to observe and why, and what you can afford. Then seek out equipment, buy it, and start. Instruments are not hard to find. Some are often readily available from garden stores and home centers. Hobby or scientific shops may have a variety. Science or school supply houses are a traditional source. And a Web search will reveal many vendors with a great range of products and prices.

Whichever way you go, enjoy making observations.

appendix c

Climate Data for Selected Stations

ASHEBORO LAT: 35.42N LONG: 079.50W ELEV: 870

	Jan	Feb	Mar	Apr	May	Jun
Min. Temp.	30.2	32.7	39.7	47	55.6	63.9
Max. Temp.	48.5	53.2	61.7	71	77.4	83.9
Mean Temp.	39.4	43	50.7	59	66.5	73.9
Precip.	4.43	3.71	4.27	3.49	4.19	3.93
HDD[a]	796	618	444	196	58	4
CDD[b]	0	0	2	16	105	271

ASHEVILLE LAT: 35.36N LONG: 082.32W ELEV: 2,240

	Jan	Feb	Mar	Apr	May	Jun
Min. Temp.	26.6	29.1	36.2	43.8	52.1	59.6
Max. Temp.	46.1	50.3	58	66.8	74.3	80.8
Mean Temp.	36.4	39.7	47.1	55.3	63.2	70.2
Precip.	3.07	3.19	3.89	3.16	3.53	3.24
HDD	888	708	555	297	126	14
CDD	0	0	0	6	71	171

AURORA LAT: 35.23N LONG: 076.47W ELEV: 20

	Jan	Feb	Mar	Apr	May	Jun
Min. Temp.	31.4	33.4	40.8	48.8	57.5	65.4
Max. Temp.	53.2	55.5	63.2	71.6	78.8	85.5
Mean Temp.	42.3	44.5	52	60.2	68.2	75.5
Precip.	4.26	3.13	4.09	3.32	4.26	4.64
HDD	704	575	406	170	34	2
CDD	0	0	2	25	131	315

Jul	Aug	Sep	Oct	Nov	Dec	Annual
68.6	67.4	61.1	49	40.1	33	49
87.7	85.6	79.6	70.3	60.7	51.5	69.3
78.2	76.5	70.4	59.7	50.4	42.3	59.2
4.12	4.26	4.22	3.59	3.16	3.26	46.63
0	0	15	202	440	705	3,478
408	357	175	36	2	0	1,372

Jul	Aug	Sep	Oct	Nov	Dec	Annual
63.5	62.4	56.4	45	36.8	29.8	45.1
84.3	82.9	76.9	67.7	57.8	49.6	66.3
73.9	72.7	66.7	56.4	47.3	39.7	55.7
2.97	3.34	3.01	2.4	2.93	2.59	37.32
0	1	46	287	530	785	4,237
277	238	96	18	0	0	877

Jul	Aug	Sep	Oct	Nov	Dec	Annual
69.5	68.5	63.6	52.4	43.2	34.9	50.8
89.1	87.7	83.3	73.8	65.4	56.8	72
79.3	78.1	73.5	63.1	54.3	45.9	61.4
5.87	6.35	4.55	3.31	2.9	3.35	50.03
0	0	4	140	333	597	2,965
443	405	259	81	13	3	1,677

BOONE	LAT: 36.13N	LONG: 081.39W	ELEV: 3,360			
	Jan	Feb	Mar	Apr	May	Jun
Min. Temp.	19.6	21.6	29.3	37.7	46.9	54.5
Max. Temp.	39.3	42.5	50	58.7	66.8	72.9
Mean Temp.	29.5	32.1	39.7	48.2	56.9	63.7
Precip.	3.97	4.14	5.18	4.7	4.87	4.58
HDD	1,101	922	786	504	269	84
CDD	0	0	0	0	17	45

CAPE HATTERAS	LAT: 35.14N	LONG: 075.37W	ELEV: 10			
	Jan	Feb	Mar	Apr	May	Jun
Min. Temp.	38.6	39	44.5	51.8	60.2	68.1
Max. Temp.	53.6	54.6	60.2	67.7	74.9	81.5
Mean Temp.	46.1	46.8	52.4	59.8	67.6	74.8
Precip.	5.84	3.94	4.95	3.29	3.92	3.82
HDD	587	518	400	187	44	3
CDD	1	1	5	29	122	297

CHARLOTTE	LAT: 35.13N	LONG: 080.57W	ELEV: 728			
	Jan	Feb	Mar	Apr	May	Jun
Min. Temp.	32.1	34.4	41.6	49.1	58.2	66.5
Max. Temp.	51.3	55.9	64.1	72.8	79.7	86.6
Mean Temp.	41.7	45.2	52.8	60.9	69	76.5
Precip.	4.00	3.55	4.39	2.95	3.66	3.42
HDD	747	585	409	180	44	3
CDD	0	1	7	40	142	323

ELIZABETH CITY	LAT: 36.16N	LONG: 076.11W	ELEV: 10			
	Jan	Feb	Mar	Apr	May	Jun
Min. Temp.	33.5	34.4	40.9	48.5	57.6	65.8
Max. Temp.	51.2	53.9	61	69.6	76.6	84.1
Mean Temp.	42.4	44.2	51	59.1	67.1	75
Precip.	4.58	3.1	4.67	3	4.43	3.73
HDD	703	584	435	196	45	2
CDD	0	0	1	17	109	300

Jul	Aug	Sep	Oct	Nov	Dec	Annual
58.8	56.6	50	37.7	29.5	22.4	38.7
76.4	75.4	70.3	61.6	52	43.6	59.1
67.6	66	60.2	49.7	40.8	33	49
4.69	4.83	3.81	3.17	4.38	3.21	51.53
28	40	159	477	728	992	6,090
108	72	14	1	0	0	257

Jul	Aug	Sep	Oct	Nov	Dec	Annual
72.9	72.3	68.5	58.8	50.3	42.6	55.6
85.4	84.8	81.1	72.6	64.8	57.3	69.9
79.2	78.6	74.8	65.7	57.6	50	62.8
4.95	6.56	5.68	5.31	4.93	4.56	57.75
0	0	2	72	244	464	2,521
440	422	297	96	24	4	1,738

Jul	Aug	Sep	Oct	Nov	Dec	Annual
70.6	69.3	63	50.9	41.8	34.9	51
90.1	88.4	82.3	72.6	62.8	54	71.7
80.3	78.9	72.7	61.7	52.3	44.4	61.4
3.79	3.72	3.83	3.66	3.36	3.18	43.51
0	0	16	165	404	655	3,208
451	405	226	43	5	1	1,644

Jul	Aug	Sep	Oct	Nov	Dec	Annual
70.4	69.4	64	52.8	43.5	36.1	51.4
87.9	86.5	81.8	72.3	63.4	55	70.3
79.2	78	72.9	62.6	53.5	45.6	60.9
4.95	4.6	5.07	2.86	2.93	3.06	46.98
0	0	5	144	356	603	3,073
440	402	242	68	10	1	1,590

FAYETTEVILLE LAT: 35.04N LONG: 078.52W ELEV: 96

	Jan	Feb	Mar	Apr	May	Jun
Min. Temp.	31.1	32.8	39.4	47	56.2	65.2
Max. Temp.	52.3	56.1	64.2	73.3	80.2	87
Mean Temp.	41.7	44.5	51.8	60.2	68.2	76.1
Precip.	4.16	3.43	4.38	3.06	3.29	4.18
HDD	722	576	412	169	38	2
CDD	0	0	3	23	138	336

FRANKLIN LAT: 35.11N LONG: 083.25W ELEV: 2,170

	Jan	Feb	Mar	Apr	May	Jun
Min. Temp.	24	25.6	32.5	38.9	49	57
Max. Temp.	47.7	52.5	60.3	68.5	75.6	81.3
Mean Temp.	35.9	39.1	46.4	53.7	62.3	69.2
Precip.	5.39	4.89	5.76	4.09	4.9	4.49
HDD	903	728	578	340	138	16
CDD	0	0	0	1	54	141

GREENSBORO LAT: 36.06N LONG: 079.57W ELEV: 897

	Jan	Feb	Mar	Apr	May	Jun
Min. Temp.	28.2	30.6	37.8	45.5	54.7	63.5
Max. Temp.	47.2	51.7	60.3	69.7	76.9	83.8
Mean Temp.	37.7	41.2	49.1	57.6	65.8	73.6
Precip.	3.54	3.1	3.85	3.43	3.95	3.53
HDD	851	679	501	245	77	8
CDD	0	0	4	25	97	263

GREENVILLE LAT: 35.38N LONG: 077.24W ELEV: 32

	Jan	Feb	Mar	Apr	May	Jun
Min. Temp.	31.3	33.5	40.3	48.3	57.3	65.5
Max. Temp.	51.6	55.2	63.3	72.4	79.3	85.7
Mean Temp.	41.5	44.4	51.8	60.4	68.3	75.6
Precip.	4.43	3.45	4.07	3.19	4.05	4.38
HDD	730	578	410	162	38	1
CDD	0	0	1	23	140	319

Jul	Aug	Sep	Oct	Nov	Dec	Annual
70.4	68.9	62.6	49.4	40.7	33.8	49.8
90.4	88.5	83.3	73.9	65	55.5	72.5
80.4	78.7	73	61.7	52.9	44.7	61.2
5.21	5.21	4.78	3.05	2.85	3.18	46.78
0	0	7	166	373	632	3,097
479	424	247	62	9	0	1,721

Jul	Aug	Sep	Oct	Nov	Dec	Annual
61.7	61.2	55	41.5	32.6	26.2	42.1
84.5	83.2	77.9	69.4	59.5	50.4	67.6
73.1	72.2	66.5	55.5	46.1	38.3	54.9
4.09	4.48	3.94	3.29	4.58	4.57	54.47
2	0	48	308	569	827	4,457
252	222	93	12	0	0	775

Jul	Aug	Sep	Oct	Nov	Dec	Annual
68.1	66.8	60.1	47.5	38.6	31.4	47.7
87.6	85.7	79.4	69.6	59.9	50.6	68.5
77.9	76.2	69.8	58.5	49.2	41	58.2
4.44	3.71	4.29	3.27	2.96	3.06	43.14
0	1	32	232	480	742	3,848
398	345	172	24	3	1	1,332

Jul	Aug	Sep	Oct	Nov	Dec	Annual
70.2	68.6	62.5	49.2	40.7	33.8	50.1
89.1	87.4	82.4	73.1	64.6	55.4	71.6
79.7	78	72.5	61.2	52.7	44.6	60.9
5.2	5.89	5.39	3.27	2.79	3.23	49.34
0	0	6	177	378	632	3,112
455	402	230	58	8	0	1,636

KINSTON LAT: 35.13N LONG: 077.32W ELEV: 55

	Jan	Feb	Mar	Apr	May	Jun
Min. Temp.	30.8	32.3	39.5	47.3	56.6	65
Max. Temp.	51.9	55.5	63.4	72.3	79	85.3
Mean Temp.	41.4	43.9	51.5	59.8	67.8	75.2
Precip.	4.44	3.42	4.26	3.44	4.1	4.64
HDD	734	591	423	177	38	2
CDD	0	0	2	20	125	306

MANTEO LAT: 35.55N LONG: 075.42W ELEV: 13

	Jan	Feb	Mar	Apr	May	Jun
Min. Temp.	35.8	37	43	51.2	59.5	68.2
Max. Temp.	51.5	53.6	60	68.7	75.8	82.5
Mean Temp.	43.7	45.3	51.5	60	67.7	75.4
Precip.	4.67	3.22	4.62	3.13	4.21	4.77
HDD	661	552	419	171	33	1
CDD	0	0	1	19	115	311

MURPHY LAT: 35.07N LONG: 084.00W ELEV: 1,640

	Jan	Feb	Mar	Apr	May	Jun
Min. Temp.	25.7	27.6	33.8	40.6	49.8	58.1
Max. Temp.	48	52.4	60.5	69.6	76.4	82.8
Mean Temp.	36.9	40	47.2	55.1	63.1	70.5
Precip.	5.81	5.07	5.86	4.58	4.85	4.76
HDD	873	700	553	305	127	11
CDD	0	0	0	7	67	175

NEW BERN LAT: 35.04N LONG: 077.03W ELEV: 16

	Jan	Feb	Mar	Apr	May	Jun
Min. Temp.	33.9	35.5	42.1	49.7	58.7	66.5
Max. Temp.	54.4	57.4	64.3	72.4	79	84.9
Mean Temp.	44.2	46.5	53.2	61.1	68.9	75.7
Precip.	4.77	3.8	4.49	3.4	4.19	4.8
HDD	651	519	370	143	29	1
CDD	6	0	5	24	148	322

Jul	Aug	Sep	Oct	Nov	Dec	Annual
69.4	67.3	61.3	48.4	40.5	33.2	49.3
88.5	86.9	82.2	73.2	64.7	55.3	71.5
79	77.1	71.8	60.8	52.6	44.3	60.4
5.88	5.62	5.45	3.51	2.8	3.56	51.12
0	0	10	183	380	644	3,182
434	375	213	53	8	0	1,536

Jul	Aug	Sep	Oct	Nov	Dec	Annual
72.3	71.6	67.4	56.8	48.1	40	54.2
86.4	85.1	80.6	71.4	63.4	55.3	69.5
79.4	78.4	74	64.1	55.8	47.7	61.9
5.25	5.64	4.93	4.1	3.4	3.67	51.61
0	0	2	112	299	539	2,789
445	415	271	84	21	1	1,683

Jul	Aug	Sep	Oct	Nov	Dec	Annual
62.8	62	56	42.8	34.3	27.9	43.5
86.1	85.5	80.2	71	60.6	51.4	68.7
74.5	73.8	68.1	56.9	47.5	39.7	56.1
4.94	4.66	3.92	3.13	4.57	4.91	57.06
0	0	36	278	529	786	4,198
294	270	129	26	1	0	969

Jul	Aug	Sep	Oct	Nov	Dec	Annual
71.1	70.1	65.1	53.1	43.7	36.3	52.2
88.3	87	82.8	74.4	66	57.7	72.4
79.7	78.6	74	63.8	54.9	47	62.3
6.48	6.84	5.45	3.39	3.23	3.84	54.68
0	0	3	125	317	564	2,722
457	419	271	87	13	7	1,759

NEW RIVER LAT: 34.42N LONG: 077.23W ELEV: 16

	Jan	Feb	Mar	Apr	May	Jun
Min. Temp.	33.9	35.6	42	49.4	58.2	66.2
Max. Temp.	55.5	58.6	65.8	73.8	80.4	86.3
Mean Temp.	44.7	47.1	53.9	61.6	69.3	76.3
Precip.	4.5	3.59	4.03	3.06	3.85	4.88
HDD	638	503	357	135	25	2
CDD	8	2	12	33	158	339

RALEIGH-DURHAM AIRPORT LAT: 35.52N LONG: 078.47W ELEV: 426

	Jan	Feb	Mar	Apr	May	Jun
Min. Temp.	29.6	31.9	38.9	46.4	55.3	63.8
Max. Temp.	49.8	54	62.5	71.8	78.7	85.5
Mean Temp.	39.7	43	50.7	59.1	67	74.7
Precip.	4.02	3.47	4.03	2.8	3.79	3.42
HDD	783	627	456	214	61	5
CDD	0	1	9	38	119	293

SOUTHPORT LAT: 34.00N LONG: 078.01W ELEV: 20

	Jan	Feb	Mar	Apr	May	Jun
Min. Temp.	33.5	35.1	41.6	48.7	57	65.4
Max. Temp.	56.4	58.3	64.5	71.4	78.1	84.3
Mean Temp.	45	46.7	53.1	60.1	67.6	74.9
Precip.	5.28	4.18	4.47	3.08	4.15	5.04
HDD	627	513	376	172	42	2
CDD	5	0	4	24	121	298

WILMINGTON LAT: 34.16N LONG: 077.54W ELEV: 30

	Jan	Feb	Mar	Apr	May	Jun
Min. Temp.	35.8	37.5	43.7	51.2	59.8	67.6
Max. Temp.	56.3	59.5	66.2	74.1	80.6	86.4
Mean Temp.	46.1	48.5	55	62.7	70.2	77
Precip.	4.52	3.66	4.22	2.94	4.4	5.36
HDD	589	474	331	134	28	1
CDD	3	4	17	65	187	361

Jul	Aug	Sep	Oct	Nov	Dec	Annual
70.8	69.7	64.4	52.2	43.2	36.1	51.8
89.5	87.9	83.7	75	67.3	59	73.6
80.2	78.8	74.1	63.6	55.3	47.6	62.7
7.08	6.47	6.27	3.32	3.33	3.69	54.07
0	0	4	139	310	543	2,656
470	428	275	95	17	3	1,840

Jul	Aug	Sep	Oct	Nov	Dec	Annual
68.5	67.2	61	48.2	39.5	32.6	48.6
89.1	87.2	81.3	71.8	62.4	53.3	70.6
78.8	77.2	71.2	60	51	43	59.6
4.29	3.78	4.26	3.18	2.97	3.04	43.05
0	1	20	194	425	679	3,465
429	379	206	39	6	2	1,521

Jul	Aug	Sep	Oct	Nov	Dec	Annual
70.1	68.5	62.9	50.9	43.3	35.8	51.1
88	87.4	83.5	75.7	68.4	59.5	73
79.1	78	73.2	63.3	55.9	47.7	62.1
6.69	7.66	8.93	3.87	3.45	4.19	60.99
0	0	6	135	300	546	2,719
435	401	252	83	24	8	1,655

Jul	Aug	Sep	Oct	Nov	Dec	Annual
72.3	71	65.9	53.9	45.1	38.1	53.5
89.9	88.3	84.1	75.6	67.8	59.6	74
81.1	79.7	75	64.8	56.5	48.9	63.8
7.62	7.31	6.79	3.21	3.26	3.78	57.07
0	0	3	95	277	497	2,429
501	455	304	90	25	5	2,017

WILSON LAT: 35.42N LONG: 077.57W ELEV: 110

	Jan	Feb	Mar	Apr	May	Jun
Min. Temp.	29.5	31.6	38.7	46.6	55.5	63.7
Max. Temp.	51.2	54.9	63	72.5	79.7	86.9
Mean Temp.	40.4	43.3	50.9	59.6	67.6	75.3
Precip.	4.31	3.42	4.4	3.12	4.08	3.82
HDD	765	610	440	184	45	3
CDD	0	0	1	20	125	312

Note: Data obtained from the National Climatic Data Center. All values refer to averages for the 1971–2000 period. For more data, see <http://www5.ncdc.noaa.gov/climatenormals/clim81/NCnorm.pdf>.

[a]HDD = heating degree days

[b]CDD = cooling degree days

Jul	Aug	Sep	Oct	Nov	Dec	Annual
68.2	66.6	60.3	47.6	39.3	32.5	48.3
90.2	88.6	83.3	73.4	64.3	54.8	71.9
79.2	77.6	71.8	60.5	51.8	43.7	60.1
5.22	4.46	4.93	3.05	3.05	3.32	47.18
0	0	14	200	405	662	3,328
440	390	217	61	9	0	1,575

Twentieth-Century Hurricanes Influencing North Carolina

YEAR	MONTH	DAY	CATEGORY[a]	NAME[b]	TRACK
1901	July	11	1	—	W across HAT[c]
1903	Sept	15	1	—	E of HAT
1904	Sept	14	1	—	SC border–Norfolk
	Nov	13	3	—	across HAT
1906	Sept	17	3	—	E–W across n SC
1908	July	30	1	—	E of HAT
	Aug/Sept	31/1	1	—	E of HAT
1910	Oct	19–20	1	—	NE E of HAT
1913	Sept	3	1	—	New Bern–Brevard
1916	July	9–10	TS	—	E Tenn
		14–16	TS	—	SW–NE through Asheville
		19	1	—	E of HAT
	Sept	6	TS	—	Southport–Roanoke Rapids
1920	Sept	12	1	—	Jacksonville
1924	Aug	25	1	—	E of HAT
1925	Dec	1	1	—	New Bern–Norfolk
1928	Sept	18–19	1	—	inland parallel to coast
1929	Oct	1–2	1	—	along fall line
1930	Sept	12	1	—	E of HAT
1933	Aug	22–23	2	—	HAT–Gatesville
	Sept	15–16	3	—	New Bern–Duck
1934	Sept	8	1	—	E of HAT
1935	Sept	5–6	TS	—	through center
1936	Sept	18	2	—	over HAT

YEAR	MONTH	DAY	CATEGORY[a]	NAME[b]	TRACK
1938	Sept	21	1	—	offshore
1940	Aug	11–17	1	—	W over northern NC
1944	Aug	1	1	—	Southport–Roanoke Rapids
	Sept	14	3	—	E of HAT
	Oct	20	TS	—	Southport–Norfolk
1945	June	25	1	—	HAT
	Sept	17	1	—	S–N through center
1946	July	6	TS	—	Wilmington–NE NC
	Oct	9	XT/NR	—	Charlotte–Norfolk
1947	Oct	12–15	1	—	E–W across SC
1949	Aug	24	1	—	E of HAT
		28	TS	—	Charlotte–W-S
1952	Aug	31	TS	Able	central
1953	Aug	13	1	Barbara	New Bern–Duck
1954	Aug	30	2	Carol	off HAT
	Sept	10	1	Edna	off HAT
	Oct	15	4	Hazel	Southport–Roanoke Rapids
1955	Aug	12	3	Connie	New Bern–Duck
		17	2	Diane	Wilmington–Raleigh–Danville
	Sept	19	3	Ione	New Bern–Duck
1956	Sept	26–27	XT/NR	Flossy	central
1958	Sept	27	3	Helene	off HAT
1959	Sept	30	TS	Gracie	Charlotte–NW
1960	July	29	TS	Brenda	Calabash–Norfolk
	Sept	11	3	Donna	New Bern–Duck
1961	Sept	20	1	Esther	E of HAT
1962	Aug	28	TS	Alma	Atlantic Beach–Nags Head
	Oct	18–19	NR	Ella	E of HAT
1963	Oct	19–27	NR	Ginny	E of HAT
1964	Aug/Sept	29-1	NR	Cleo	Charlotte–Elizabeth City
	Sept	13	NR	Dora	E of HAT
	Sept	21–23	NR	Gladys	E of HAT
	Oct	16	1	Isbell	Morehead City–Norfolk
1966	June	11–12	1	Alma	E of HAT

YEAR	MONTH	DAY	CATEGORY[a]	NAME[b]	TRACK
1967	Sept	10–16	TS	Doria	??
1968	June	7–13	NR	Abby	Charlotte–HAT
	Oct	19–20	1	Gladys	E of HAT
1969	Sept	8	NR	Gerda	HAT
1970	May	26	TD	Alma	central NC
1971	July	5	NR	Arlene	HAT
	Aug	13–14	NR	Beth	HAT
	Aug	27	TS	Doria	Atlantic Beach–Nags Head
	Sept/Oct	1	1	Ginger	Atlantic Beach–S central VA
1972	June	20–21	TS	Agnes	Charlotte–Norfolk
	Sept	8–9	NR	Dawn	off HAT
1973	Oct	25–26	NR	Gilda	off HAT
1975	June	28	TS	Amy	off HAT
	Oct	1–2	NR	Gladys	off HAT
	Oct	26–27	TS	Hallie	off HAT
1976	Aug	9	NR	Belle	off HAT
	Aug	20–21	TS	Dottie	SE of state
1977	Sept	5–8	??	Clara	into SE from W
1978	Sept	1–2	NR	Ella	off HAT
1979	Sept	5	TS	David	through central portion
1981	Aug	20–21	TS	Dennis	off HAT
1982	June	18–19	TS	—	off HAT
1984	Sept	9–14	2	Diana	Long Beach–Oregon Inlet
	Oct	12–15	NR	Josephine	off HAT
1985	July	24–26	NR	Bob	central Carolina–W VA
	Sept	26–27	3	Gloria	crossed HAT
	Nov	22	TS	Kate	off HAT
1986	June	7–8	TS	Andrew	off HAT
	Aug	17–18	1	Charley	Outer Banks
1989	Sept	21–22	3	Hugo	Charlotte
1991	Aug	20	1	Bob	off HAT
1992	Sept	22	TS	Danielle	off HAT
1993	Aug	31	3	Emily	off HAT

YEAR	MONTH	DAY	CATEGORY[a]	NAME[b]	TRACK
1994	July	6	NR	Alberto	offshore
	Aug	31	TD	Beryl	NW GA–W VA
1995	June	6	NR	Allison	coastline
	Aug	27	NR	Jerry	Charlotte–HAT
1996	June	20	TS	Arthur	S Outer Banks
	July	12–13	2	Bertha	Topsail–Roanoke Rapids
	Sept	5–6	3	Fran	Bald Head–Danville
	Oct	8	NR	Josephine	SW–NE across state
1997	July	24	TD	Danny	Charlotte–Norfolk
1998	Aug	26	2/3	Bonnie	Wilmington–HAT
	Sept	3	NR	Earl	Laurinburg–HAT
1999	Sept	4	TS	Dennis	erratic
	Sept	16	2	Floyd	Cape Fear–Norfolk
	Oct	18	1	Irene	off HAT

[a]Numbers indicate hurricane category; TS = tropical storm; TD = tropical depression; NR = not rated (hurricane remnant).

[b]No names prior to 1950.

[c]HAT = Cape Hatteras

Sources: Jay Barnes, *North Carolina's Hurricane History*, rev.ed. (Chapel Hill: University of North Carolina Press, 1998); James D. Stevenson, "A Historical Account of Tropical Cyclones That Have Impacted North Carolina since 1586" (National Weather Service NOAA Technical Memorandum NWS ER-83, Washington, D.C., 1990); National Hurricane Center, <http://www.nhc.noaa.gov>

index